The Energy Crisis

Michael Tanzer

The Energy Crisis:
World Struggle for Power and Wealth

Monthly Review Press
New York and London

Brief portions of this work appeared previously in *The Political Economy of International Oil and the Underdeveloped Countries*, copyright © 1969 by Michael Tanzer and reprinted by permission of Beacon Press; and in *The Sick Society: An Economic Examination*, copyright © 1968, 1969, 1971 by Michael Tanzer and reprinted by permission of Holt, Rinehart and Winston, Inc.

Library of Congress Catalog in Publication Data
Tanzer, Michael.
    The energy crisis.
        1. Energy Policy. 2. Petroleum industry and trade.
3. Power resources. 4. Power (Social sciences) I. Title.
HD9502.A2T35   333.7   74-7787
ISBN 0-85345-346-2

First Printing

Monthly Review Press
62 West 14th Street, New York, N.Y. 10011
21 Theobalds Road, London WC1X 8SL

Manufactured in the United States of America

To my mother, Sara Rhoda Tanzer,
a brave and able unsung woman,
with deepest love and gratitude

Acknowledgments

I wish to express my appreciation to a number of people who helped make this book possible. First, to Harry Braverman, Harry Magdoff, and Paul Sweezy of Monthly Review, who first suggested the idea to me of doing a popular but rigorous book on the subject. Second, to Harvey O'Connor, whose generous assistance has helped me to broaden my contacts in the oil area, and whose own highly readable books on the subject have given me confidence that such a work could be done. Third, to Charles Constantinou of the United Nations Energy Section, an international civil servant in the best and highest sense, with whom I have had many illuminating discussions over almost a decade, and whose thoughtful reading and criticism of this manuscript were helpful as always. Fourth, to Susan Lowes of Monthly Review Press, who not only expeditiously moved this book through the various stages of publishing, but also made helpful substantive suggestions. The fact that all of these people are not only professional colleagues but also friends makes me particularly happy to express my heartfelt thanks.

Finally, there is the continuing debt I owe to my wife, Dr. Deborah Tanzer, and to my sons David, Kenneth, and Charles. Not only did they bear the burden of my preoccupation with this book, but their goodness and love are a daily inspiration.

Michael Tanzer
August 1974

7

Contents

Introduction

In the last few years, the so-called energy crisis has been catapulted to the forefront of the world's concern. The purpose of this book is to reveal in as clear a way as possible the true nature of this crisis. My aim is to show that there is no real energy crisis, in the sense of a physical shortage of energy resources; rather, there is an artificially contrived scarcity generated by various forces operating within the overall framework of the international capitalist economy.

Some of these forces may have fundamentally divergent goals, but in the short run at least they may cooperate for immediate self-interest. This is true despite the fact that, as I hope to show, the short-run "equilibrium" worked out by these forces is inherently unstable, and is bound to lead in the long run to disastrous results for the world. For this reason the international energy crisis may be seen as a classic example of the irrationalities of the international capitalist system.

At the present time we appear to have a temporary coalition of forces between the international oil companies, the U.S. government and the oil-producing countries. This coalition maintains the price of crude oil far above its cost of production, while allowing enormous profits for the companies and increased U.S. economic power vis-à-vis its European and Japanese rivals. At the same time, the sudden shift in flows of wealth to the oil-producing countries resulting from the recent increases in oil prices appears to be so great as to be unsustainable by many of the economies of the developed world, let alone by the impoverished, oil-importing underdeveloped countries which make up the great bulk of the Third World.

Thus, at this juncture it would appear that ultimately the most likely "solutions" to the international oil problem would be either intervention by the developed countries in the Middle East, including possibly use of armed force, to drive down the price of oil, or an international depression which by reducing sharply the demand for oil would help offset the impact of the recent oil price increases. Since the developed countries are deeply divided as to their respective self-interests vis-à-vis the oil crisis, and since military intervention in today's revolutionary world is fraught with danger, a world depression seems the more likely outcome. This is all the more so because the oil crisis interacts with and sharply worsens the unsolved international monetary crisis which has always had the potential for plunging the international economy into depression. Given the many sources of instability in the contemporary world capitalist economy, even a rollback in oil prices would seem insufficient to avoid this denouement.

In order to see why the world has reached this grim predicament we have to examine closely the major forces which operate today on the world energy scene. The most important of these are the international oil companies, their home governments, the other countries of the developed world, the oil-exporting underdeveloped countries, the Communist countries, and the oil-importing underdeveloped countries. Moreover, we have to examine the historical development of the relationships among these forces.

Thus, this book attempts to combine a topical and historical analysis, with greater emphasis on the recent history. We first need to have a good picture of the development and current status of the world's supply of and demand for energy resources. It is to this task that we turn in the first chapter.

1
World Energy Resources: Supply and Demand

1866—U.S. Revenue Commission says synthetics available if oil production should end.

1883—Little or no chance for oil in California—U.S. Geological Survey.

1891—Little or no chance for oil in Kansas or Texas—U.S. Geological Survey.

1908—Maximum future supply of oil 23 billion barrels—officials of U.S. Geological Survey.

1914—Total future production only 6 billion barrels—officials of the U.S. Bureau of Mines.

1920—United States needs foreign oil and synthetics; peak domestic production almost reached—Director of U.S. Geological Survey.

1931—Must import as much foreign oil as possible to save domestic supply—Secretary of the U.S. Department of the Interior. (East Texas field discovered in 1930 but full potential not immediately recognized.)

1939—U.S. oil supplies will last only 13 years—U.S. Department of the Interior.

1947—Sufficient oil cannot be found in the United States—Chief of Petroleum Division, U.S. Department of State.

1949—End of U.S. oil supply almost in sight—Secretary of the U.S. Department of the Interior.
From a chart prepared by the Independent Petroleum Association of America in 1952

The petroleum age began about one hundred years ago when Edwin Drake discovered oil in Pennsylvania in 1859. Until the turn of the century, oil was used primarily for kerosene for lighting and heating, and secondarily for lubricants. Around the turn of the century, fuel oil for boilers in ships and industry came to prominence as kerosene's significance faded. The automobile era in the United States, which can be conveniently dated from 1911, the year the Standard Oil cartel was legally broken up, was the biggest cause of the rise of oil, along with World War I, which proved to the hilt Clemenceau's view that "a drop of oil is worth a drop of blood." By 1929 petroleum accounted for one-third of U.S. energy consumption (different fuels are converted to statistical equivalency on the basis of latent energy content). Even so, in 1939, the eve of World War II, coal still predominated over oil and natural gas, so that these "petrofuels" together accounted for only 45 percent of total U.S. energy consumption. During and after World War II, however, military, transport, and industry demand for petrofuels skyrocketed; by 1952 they accounted for about two-thirds of total U.S. energy consumption, and this proportion has increased steadily to over three-fourths today.*

The rest of the developed world, lacking the auto boom of the United States, lagged far behind in oil consumption. Even by 1950, in Western Europe petrofuels accounted for only one-seventh of energy consumption, and in the Far East (primarily Japan), one-fifth, compared to three-fifths in the United States. However, following World War II, oil con-

* Sources will be found at the end of the book.

sumption in Western Europe and Japan rose sharply, first from heavy fuel oil replacing coal and later from an auto boom which triggered gasoline demand. By 1970 Western Europe and Japan used petrofuels for two-thirds of their energy consumption.

Finally, it is interesting to note that the centrally planned economies, i.e., the Communist countries, moved much more slowly in shifting from coal to petrofuels than did the developed capitalist countries. Starting from the same relative share as Western Europe in 1950, one-seventh, the Communist countries' petrofuel consumption increased to only one-fifth by 1960, and to less than two-fifths in 1970. This is true despite the fact that the Communist countries' total energy consumption increased more than threefold in this 1950–1970 period, while that of Western Europe only doubled. Part of this relatively slower penetration rate for petrofuels in the Communist countries is undoubtedly due to the lack of an auto boom. But even more important, as I shall discuss later, is the planned nature of their economies, which aimed at self-sufficiency, so that petrofuel consumption was restricted largely to their ability to produce these fuels.

This leads to a very interesting observation about energy consumption in the twentieth century. Up to 1950 petrofuel consumption growth was overwhelmingly centered in the United States, which as late as 1952 accounted for about 70 percent of world consumption. This was associated with the fact that the United States, as the leading producer of petrofuels, accounted for about 60 percent of world production. This close correlation between supply and demand existed partly because natural gas, which often comes out of the ground as a joint product with crude oil, could not then be transported over the long distances between the underdeveloped and the developed countries. Hence most of the natural gas produced in Venezuela and the Middle East was "flared" or burned at the production site and never utilized. In the United States, of course, natural gas was not flared but shipped by pipeline internally. Thus the United States accounted for over 90 percent of useful gas production in the world, compared to only 53 percent of world crude oil production.

Elsewhere in the world coal remained king until around 1950. Because coal resources are spread fairly widely throughout the world, and it is bulky and expensive to ship over long distances, the countries of the world were generally autarkic with regard to energy supply. Thus, even as late as 1952, the percentage of total energy consumed which was supplied by indigenous production was as follows: United States, 99 percent; Western Europe, 86 percent; Far East, 97 percent; Communist countries, 101 percent (i.e., net exports).

This is not to say that the term "international" in reference to the oil industry was a misnomer. In fact, throughout the twentieth century very important quantities of oil moved in intercountry and interregional trade, and as we shall see later, the struggle for control both of oil supplies and of oil markets was a continuing source of international tension and even war. However, the last twenty-five years have seen a qualitative change, whereby not only has King Coal been overthrown by the petrofuels, but overthrown primarily by petrofuels imported from the Third World. Thus, for the first time in modern history, major countries and regions of the world have become heavily dependent upon imported petrofuels for their vital energy needs, and by 1972 the proportion of total energy supplies provided by indigenous production was as follows: United States, 85 percent; Western Europe, 41 percent; Japan, 13 percent; Communist countries, 104 percent.

What brought about this sharp swing (for the non-Communist countries) away from self-sufficiency in energy resources? A number of interrelated factors can be cited.

First and most obvious is the finding of enormous low-cost crude oil reserves in the Middle East in the late 1930s and early 1940s, reserves which were developed rapidly after World War II. Proven reserves in the Middle East were estimated to have jumped from only six billion barrels in 1940 to 300-500 billion today, or from one-sixth of the world's reserves in 1940 to 50-60 percent today. The average cost of

production of these Middle Eastern reserves has been estimated to be about ten cents per barrel, compared to average
production costs in Venezuela of close to fifty cents and in
the United States of $1.25 per barrel.

The second factor was the control of the non-Communist
world's crude oil supplies by a handful of enormously powerful, vertically integrated international oil companies. The
"Big Seven" consisted of five American companies—Exxon,
Texaco, Standard of California, Mobil, and Gulf; one British
company, British Petroleum; and one mixed British-Dutch
company, Royal Dutch Shell (sometimes France's Compagnie
Française des Pétroles—CFP—is considered the junior member
of a Big Eight).* In 1949 the Big Seven controlled 69 percent
of the non-Communist world's oil reserves, as well as 57 percent of its refining facilities.

While the key role of the major international oil companies
will be discussed in detail in later chapters, the main point to
be noted here, regarding their ability to swing entire nations
and regions away from self-sufficiency in energy resources, is
that within their home countries they wielded enormous economic and political power. Thus, their home governments,
and particularly the U.S. government, were willing and eager
to go to great lengths to advance the interests of these companies. (In addition, there was a basic symbiosis between the
oil companies and their home governments which we shall
discuss later.) When taken in conjunction with the third factor, this power proved decisive in the swing to reliance on
imported petrofuels.

This third factor was the U.S. rise to complete hegemony
in the capitalist world following World War II, a war which
had badly weakened the former leading oil powers—Great
Britain, France, and the Netherlands—and also destroyed the
leading Axis powers—Germany, Japan, and Italy. Since war-
torn Europe and Japan were heavily dependent upon U.S.
assistance for reconstruction, the oil companies and the U.S.

---

* In this book I have generally used the present corporate name for
companies which have had different names in earlier historical periods.

government used this opportunity to virtually ram American-controlled oil down the throats of the world to replace coal.

Thus, Walter Levy, head of the Marshall Plan's oil division, and previously an economist for Mobil, noted in 1949 that "without ECA [the Marshall Plan] American oil business in Europe would already have been shot to pieces," and commented that "ECA does not believe that Europe should save dollars or earn foreign exchange by driving American oil from the European market." Some $2 billion of total Marshall Plan assistance of $13 billion was for oil imports, while the Marshall Plan blocked projects for European crude oil production and helped American oil companies to gain control of Europe's refineries. All this was done without regard for the effects on domestic employment in coal, loss of internal self-sufficiency, or balance of payments considerations.

This process was helped along by the fact that, because oil from the Middle East and Venezuela was so low in cost, the international oil companies could easily set a price for fuel oil which would insure that European and Japanese industrial consumers would favor it over coal. Such cutthroat pricing was further aided by the consideration that the oil industry was traditionally oriented to maximizing production and prices of gasoline for the U.S. market, for which the consumer had no real alternative. Hence, virtually any price which could be obtained for the other products derived from a barrel of crude oil, especially heavy fuel oil for industry and light fuel oil for home heating, would be gravy for the oil companies. Finally, the fact that European and Japanese industry had to compete with American firms which were switching to oil meant that the governments of Europe and Japan could not realistically block the increasing reliance on foreign energy supplies, even if U.S. government aid and pressure were not a factor. As a result, in the words of Taki Rifai, author of a recent comprehensive study on the pricing of crude oil:

> The price of heavy fuel oil, as set by the majors in Western Europe, happened to be quite low indeed, since it was cheaper than alternative sources of primary energy, mainly coal. . . . It

appears as if the low price set by the majors for heavy fuel oil was to ensure market penetration and control. This process was further reinforced by political considerations. Since European reconstruction was financed by American dollars, it would be advisable to invest in enlarging the outlets for fuel oil largely derived from dollar crude rather than in reinforcing and reviving local coal industries.

Whatever the reasons, we have now swiftly reached an unprecedented situation where, as we have seen, the leading developed countries, excluding the United States, as well as most of the underdeveloped countries, have become overwhelmingly dependent upon foreign crude oil for their vital energy supplies; even the United States has become significantly dependent on foreign sources. It is important to realize that this crucial development in the capitalist world did *not* take place because of lack of energy resources within the various countries.

This brings us to the very crucial and badly misunderstood question of the world's reserves of various energy sources. It is a truism of economics that the concept of energy (or mineral) "reserves" is an elastic one, depending partly on the state of technology, which affects the cost of production. For example, if new technology allows mining of low-quality coal which was formerly uneconomic to produce, then the estimated "reserves" of coal can be increased. Hence, most estimates of reserves assume some price, usually the current one, as a "cut-off": if the resource cannot be produced at a profit at that price, it is not counted in reserves.

Another point to keep in mind is that estimates of energy reserves are always tentative, because much of the world's terrain has never been fully explored. When there is a great surplus of a commodity, little effort will be made to look for more, even though, as oil experts put it, "Oil is where you find it." For example, there is probably far more oil in the Middle East than now estimated; but once the international oil companies found the presently estimated reserves, which are far greater than they could extract over the likely lifetime of their concessions, they had no incentive to explore further within the region. Finally, it must be recognized that it is in the self-interest of the private companies which control most energy resources in the non-Communist world to paint a picture of scarcity in order to keep prices as high as possible. (The ability of the companies to do so was graphically illustrated when a top oil company official told the director of a U.S. government task force on oil that "he regretted having given us the optimistic-pessimistic data as distinct from the pessimistic-pessimistic data.")

Thus, if one's real concern is that, as some claim, the world is running out of energy resources in an absolute sense, the available data should be quite reassuring. (Even the question of whether we are running out of low-cost energy resources is still an open one.) A recent comprehensive study by Yale professor William Nordhaus presents the following picture of world reserves. Recoverable coal reserves have been estimated at over six trillion tons, or enough to fill all of the world's energy needs, at today's consumption rates, for well over 500 years. Total recoverable petroleum reserves have been estimated at more than 200 billion tons, or over seventy times current annual consumption. Similarly, the world's estimated reserves of natural gas are over 150 trillion cubic meters, or more than 150 years worth at today's consumption rates. The world's hitherto largely unexploited oil shale reserves are believed to be on the order of 200 times larger than conventional oil reserves. Finally, with the breeder reactor, which is expected to be perfected before the end of the century, estimated uranium reserves would contain more than a million times the energy of all fossil fuels. At the end of the line, fusion of hydrogen, as in the hydrogen bomb, would provide virtually unlimited energy. And, of course, none of this takes into account the vast possibilities of nonpolluting energy sources such as the sun, winds, tides, and hot springs, use of which may be preferable for environmental reasons.

It is true that the world's known energy resources tend to be unequally distributed relative to population. However, despite the fact that around three-fourths of the world's estimated coal reserves are in the United States, the Soviet

Union, and China, coal deposits are sufficiently widespread to have provided indigenous energy resources in many other countries. Thus, indigenous coal reserves have been estimated to be sufficient to cover today's total energy consumption for 500 years in the United States, 3,000 years in the Soviet Union, and 2,000 years in China. Even in Western Europe, estimated coal reserves as a multiple of annual total energy consumption were eighty for West Germany and fifty for Great Britain. Again, in the underdeveloped countries, India has 600 years of reserves of coal to cover total consumption needs; relatively little oil has been discovered thus far, yet petrofuels now account for close to one-third of total energy consumption there.

In conclusion, despite the existence of widespread large-scale coal reserves, the non-Communist countries have in the last quarter-century become increasingly dependent upon imported petrofuels for their very lifeblood. This "gamble" was easier to take because for most of this period fuel oil prices were cheap relative to coal. Since the 1970s, however, when oil prices first increased rapidly and then skyrocketed after the 1973 October War in the Middle East, the risks of this gamble have begun to be felt in many countries. From 1969 to 1974, costs of importing oil have jumped from under $20 billion to over $100 billion.

These enormous increases have been paid primarily to the oil-producing countries and secondarily to the international oil companies. However, these massive financial shifts raise two crucial and related problems that now form the core of the real "energy crisis." First, how will the importing world be able to pay for this oil in the future? And second, what might be done with the vast foreign exchange reserves that the oil-producing countries could amass, theoretically at least, as a result of these payments?

In a later chapter we will explore in detail the possible answers to these $64 billion questions. But before we can assess the enormous significance of these incomprehensible numbers to all the peoples of the world, we must first look at the major actors on the world energy stage. Only then can we begin to see the possible outcome of events, and to see what might be an alternative solution to the energy crisis in a more rational world.

2
The International Oil Companies: From America to Zambia

Not so long ago I was a guest lecturer at a United Nations–sponsored conference in New Delhi, where leading oil experts and delegates from thirty countries assembled for two weeks to discuss various aspects of the oil industry in underdeveloped countries. In the course of a sometimes heated debate between one leading consultant and myself, and in apparent exasperation at my repeated criticisms of the international oil companies, he finally asked me, somewhat petulantly: "Well, what do you want the companies to do?" When after a brief pause I suggested that the best thing for them to do would be to "disappear" from the world scene, a sudden hush descended over the conference. It was as if I had stood up in St. Peter's Cathedral in Rome and cried out, "God should die!"

Yes, so deeply and widely are the international oil companies entrenched in the capitalist world that it is hard even for experts to remember that a century ago they did not exist, and are likely in far less time than that to disappear in the future. To see how these companies have become such venerable giants today it is useful to briefly review their pattern of growth in the last hundred years.

As Harvey O'Connor, an astute observer of the industry, has noted:

> For a half-century the history of oil was also the personal history of John D. Rockefeller, who tamed an anarchic industry and brought it under the direct control of Standard Oil. The oft-told tale ran the spectrum of the devices of monopoly. Competitors were bought out or ruined, legislators and public officials were also bought out (and many ruined), laws were flouted with impunity or by stealthy indirection.

Rockefeller made his first investment in oil in 1862, and by 1870 had formed Standard Oil. For many years Standard Oil concentrated on building a monopoly in refining and transporting and marketing oil, largely leaving the riskier area of exploration and production to smaller entrepreneurs. At the same time it retained a virtual monopoly in world markets. This monopoly was undermined by two sets of events. In the United States market the critical date was 1901, when the Spindletop field in Texas was discovered—the largest producer yet found on the American continent, with 100,000 barrels a day gushing from the initial well. Out of the need to find capital for developing and marketing this bonanza, the original finders turned to Andrew Mellon, banker and industrialist. Mellon's Gulf Oil Corporation was formed to handle this transaction. Other competitors for Standard Oil arose out of the Spindletop field as the Texas Corporation (Texaco) got leases there and Shell Transport and Trading Company of London undertook to market some of the oil, largely to the British navy.

Outside the United States the Standard Oil monopoly was challenged by oil discoveries in Russia and the Dutch East Indies. The Nobels and the Rothschilds developed Russian oil into a formidable competitor to Standard Oil in Europe by the late 1880s; Standard fought back by using its monopoly of American kerosene to build its markets abroad, so that by the end of the nineteenth century Standard had two-thirds of the British market and four-fifths of the continental European market, all of Latin America and Mexico, virtually all of the Far East, and three-fifths of Canada. Oil was found in 1885 in the Dutch East Indies; Royal Dutch Petroleum was organized in 1890 and began competing fiercely with Standard for the fabled "oil for the lamps of China" market. Worldwide competition developed further when Royal Dutch merged in 1907 with Shell Transport and Trading, which had Russian and American oil supplies, to form the Royal Dutch Shell Group.

In 1911 the Standard Oil monopoly was broken up by the antitrust action of the U.S. government, and the company was divided into a number of major oil companies, among which the most important were Standard Oil of New Jersey (now Exxon); Standard Oil of New York (now Mobil); Standard Oil of California; Standard Oil of Ohio; and Standard Oil of Indiana. The first three of these Standards, along with Gulf, Texaco, Royal Dutch Shell, and British Petroleum (which was formed as Anglo-Persian in the early 1900s to operate a British oil concession in Iran), comprised the seven companies which came to be known as the "Seven Sisters" or the Cartel. For the last twenty-five years the Seven—five American companies, one British, and one mixed British-Dutch—have dominated the international oil industry, for reasons that we will come to later.

Within the United States, fifteen or so smaller oil companies, specialized either by region or function, grew to considerable size after the 1911 split-up of Standard Oil. These included Atlantic Company (now part of Atlantic Richfield), Tide Water (now part of Getty), Continental, Union, Phillips, Sinclair (now part of Atlantic Richfield), and Cities Service. The distinguishing characteristic of these firms is that, while large, they were all essentially domestic operations, at least until after World War II; they fattened off the huge American domestic market and were largely insulated from the world oil scene.

Up until 1928 the history of the international oil industry was a continuous series of fierce struggles for markets and profits, primarily between Standard Oil of New Jersey and Royal Dutch Shell. As Harvey O'Connor notes: "In the mid-1920s Standard and Royal Dutch were still the prime antagonists in the oil world. Standard was still fighting for entry into the rich Iraq fields, maneuvering with Royal Dutch for access to the Soviet fields, protecting its Mexican properties against rising nationalism, angling for concessions, and extending markets around the globe." Every conceivable tactic was used, from price wars and bribery to violence. As the official historian of Standard Oil admitted, "for its unsavory reputation, the world petroleum industry could only blame its leaders, fighting with all the means at their command—the

worse as well as the better—to channel the forces of militant
nationalism and to restore an economic system which was
not to be restored." Moreover, it should be noted that as the
oil demand boomed not only for automobiles, but for fueling
ships, planes, and tanks, the struggles for position in the
international oil industry increasingly became intertwined
with national rivalries. This put oil smack in the middle of
the constant jockeying for military and economic power
which characterized relationships among the developed capitalist countries. This will be examined more closely in the
next chapter.

The international "oil war" ended, just like World War I,
with a formal armistice agreement, negotiated in 1928 at
Achnacarry, Scotland, by the heads of Standard Oil of New
Jersey, Royal Dutch Shell, and British Petroleum. The "as
is" agreement, as it was called, put an end to competition by
ruling out new production facilities until the old ones were
being used at capacity, and by fixing market shares of the Big
Three at the 1928 level. Price competition was to be elimi-
nated by having each market supplied from the nearest
source at a world price based on the high-cost Texas Gulf
area, plus about one dollar per barrel profit; this assured the
big international oil companies superprofits from their lower-
cost non-U.S. oil. It might be noted that this oil has been
relatively low-cost because oil fields outside the United States
tend to be much larger, thereby requiring fewer wells to be
drilled per barrel of oil produced. In addition, the United
States has been vastly overdrilled owing to the irrationalities
of its tax laws and wasteful competition in domestic oil exploration.

By the late 1940s, when the international oil industry was
poised for its fantastic expansion, the Big Three had effectively grown to the Big Seven. Standard Oil of California
(Socal), not being party to the Big Three agreement, in 1936
obtained oil concessions in Bahrein, and far more important,
the fabulous concession which opened up to it all of Saudi
Arabia's oil. Socal joined with Texaco, which had
strong marketing capacity east of Suez, to set up the Arabian

American Oil Company (Aramco), which until now has exploited virtually all of Saudi Arabia's oil. Mobil gained entry
into the big time through receiving a 12 percent interest in
Iraq's petroleum in 1928, as well as a 10 percent interest in
Aramco in 1946 (at the same time that Standard of New
Jersey bought a 30 percent interest in Aramco). Finally,
Gulf, which had oil in Venezuela, in 1934 got a concession
(in 50–50 partnership with British Petroleum) for all of Kuwait's oil, which in the post–World War II period turned out
to be the key to its fortune.

By 1949 the Big Seven accounted for 55 percent of the
non-Communist world's total crude oil production, 57 percent of its refining capacity, and dominated transport and
marketing as well. And this crude oil production figure
underestimated the Seven's real power and profitability, since
they accounted for less than one-third of the high-cost U.S.
crude oil, but 88 percent of all of the rest of the world's
production, where the larger profits were made.

In terms of crude oil production Exxon was at the top of
the tower, trailed by Royal Dutch Shell, British Petroleum,
Gulf, Texaco, Socal, and Mobil; in 1949, Exxon's worldwide
production was over 50 percent greater than Royal Dutch
Shell's and four times greater than Mobil's. In terms of
profits, the rankings were similar, except that British Petroleum, which suffers from a lack of refining and marketing
facilities, was the least profitable of the Seven.

It is interesting to note that after two decades of tumul-
tuous growth, during which the Big Seven's crude production
increased almost fivefold, the relative positions of the members had changed very little. In 1972 Exxon was still the
kingpin, and Mobil was still last with less than one-third of
Exxon's production. In terms, again, of profits, Exxon was
still at the top, while British Petroleum brought up the rear.

The key to this relative lack of change in market positions
was the phenomenal growth of the Third World, and the
Middle East in particular, as the key source of the world's oil,
and the relative decline of the United States as an oil producer. Back in 1949, when the non-Communist world's

petroleum reserves were estimated at seventy-four billion barrels, the United States was believed to have two-fifths of this and the Middle East a similar amount. Today (1973), when estimates of the world's reserves are about ten times greater than in 1949, the Middle East is credited with perhaps 50-60 percent, while the U.S. share has dropped to 5 percent of the total. During this period the Middle East's share of world production has risen from about 15 percent in 1949 (up from only 6 percent ten years earlier) to 38 percent in 1973.

In terms of both size and profitability, the international oil companies have always been among the most significant concentrations of economic power within the capitalist world. Thus, in the United States, back in 1947, three of the Big Five American companies (Exxon, Mobil, and Texaco) ranked among the top ten U.S. manufacturing firms (in terms of total assets). So rapid has been the growth of the Big Five that now, twenty-five years later, all of them rank in the top ten! Their total assets (excluding depletion and depreciation) at the end of 1973 amounted to $69 billion, or one-eighth of the total assets of *Fortune*'s top 500 manufacturing companies. Even more significant, the Big Five's profits in 1972, a year before the enormous price jump which increased their incomes by two-thirds, amounted to $3.8 billion, or almost one-seventh of *Fortune*'s total manufacturing profits. In 1973, the Big Five's profits were $6.2 billion, or one-sixth of the top 500's manufacturing profits.

The vast expansion of the world's oil business in this period allowed rapid growth for numerous other oil firms besides the Big Five. Thus, while in 1947 sixteen out of the top one hundred manufacturing firms in the United States were oil companies, by 1972 the number was twenty. The smaller fifteen of these twenty had total assets in 1972 of $45 billion, or three-quarters that of the Big Five, but net income of $2.1 billion, or little more than half that of the Big Five. This relatively lower profitability for the "fifteen minors" reflected the fact that while most of them had made the transition after World War II from wholly domestic companies to at least partially international ones, the Big Five still had the advantage of having monopolized most of the world's low-cost crude oil in the Middle East.

At this point a word should be said about the historical development of the international oil companies in terms of the changing scope of these companies' operations. The decision to expand this scope qualitatively is normally due to a combination of two factors: a desire to find profitable investment outlets for current revenues, and/or a desire to reduce the instability or riskiness of future profits. The behavior of the international oil companies as they grew is a classic case of this pattern.

In the beginning, as we have seen, the first major company, Standard Oil, was not completely "vertically integrated," as it preferred to develop its monopoly in refining, transporting, and marketing, leaving the risky area of crude oil exploration and production to others. Such a strategy made sense as long as others could find sufficient oil and Standard could retain its position as a monopoly refiner and marketer. However, when the Spindletop field blew in in 1901, bringing vast quantities of crude oil under the control of recently founded Gulf Oil, Standard Oil refused to make a deal, and Gulf was forced to break new ground by setting up a wholly vertically integrated operation.

Thus, Gulf operated at every level from crude oil exploration through marketing. Similarly, other independent companies which were to find large quantities of crude oil had to integrate forward into refining and marketing in order to have outlets for the crude, particularly at times when there was a glut of crude. On the other hand, independent marketers and refiners found it necessary to integrate backward into crude oil production in order to have assured supplies, particularly at times when there was a scarcity of crude. Thus, vertical integration of various kinds became the rule in the industry so that today virtually all large oil companies operate at every level of the industry.

Because the discovery of large fields of crude oil has always been a somewhat random event, from time to time the industry has been hit by "shock waves" of vast oil finds which cause great upheaval. One of these was the Spindletop field which helped break the Standard Oil monopoly. Another was the 1930s' discovery of the vast East Texas oil field, which drove oil prices down to ten cents per barrel and led to the internal prorationing laws which restricted competition within the United States in order to keep crude oil prices high. The discovery of vast oil fields in the Middle East since the 1930s was the third major event which helped to propel most of the major U.S. oil companies into a further expansion of their scope—changing from primarily if not totally domestic companies into significant if not major international companies. Here again the impetus was to gain access to low-cost crude oil, in order both to enjoy high profits when oil prices were high and to be able to compete when oil prices were low.

The influence of the Middle Eastern oil fields in shifting most U.S. oil companies from domestic to international scope can hardly be overstated. Until the post–World War II era, only Exxon, which in 1939 produced as much crude oil as the rest of the Big Five combined, was a truly international company. It drew two-thirds of its crude oil production from abroad, while all other American companies had relatively little foreign production—e.g., for Mobil, less than one-fifth.

In 1972, while Exxon drew four-fifths of its crude production from outside the United States, and about two-fifths from the Middle East, Standard of California got two-thirds of its production from the Middle East, Gulf three-fifths, and Texaco and Mobil about one-half each. Even a formerly wholly domestic company like Standard Oil of Indiana drew two-fifths of its crude oil from overseas, with one-fifth coming from the Middle East. Moreover, since (until the October War in 1973) Saudi Arabian oil, controlled by four of the Big Five American companies (excluding Gulf), was slated to be the biggest single growth source for the world, it is clear that the Middle East was to become increasingly important for the American companies. And of course, since Middle Eastern crude oil is the world's cheapest, these proportions understate the contribution of Middle East crude to each company's crude oil profits.

The most recent expansion in scope of the international oil companies has been the move into oil-related areas of operation. In the 1950s and early 1960s the companies moved heavily into the petrochemical business. This was a natural step, since they own the basic feedstocks and have large quantities of the necessary capital required to invest, and also since the chemical industry uses processing patterns much like those they were familiar with from petroleum refining. As a result, by 1962 the oil companies owned or operated over one-third of all U.S. petrochemical plants; petrochemicals accounted for over 60 percent of the value of all organic chemicals produced in the United States.

Far more significant, however, was the next wave of expansion, beginning in the early 1960s, when the major oil companies began entering into competitive fuel industries, particularly coal and nuclear power. Their mixed success in earlier petrochemical ventures had reinforced the historic lesson that the companies' greatest advantage was to be gained from becoming monopolistic suppliers of raw materials; it is in such industries that one can get the highest monopoly profits, owing to the fact that control of raw materials can bar new entry into an industry, thereby allowing very high rates of return on investment. Thus, the enormous profits of the Big Five in the 1950s derived primarily from their control over low-cost crude oil in the Middle East and Venezuela. In my view, the lesson of this experience, combined with certain key changes in the international oil industry in the 1960s, was what led the big oil companies to seek monopolistic control over all energy sources, particularly within the United States.

Before discussing this overall strategy in more detail, we should note the three sets of events in the last decade which led the companies to this strategy. First, the 1960s saw declining profit rates in the international oil industry, partly because increased competition triggered by newcomers drove down the price of crude oil. Second, the growth of environmental problems, particularly in the United States but also increasingly in Japan and Western Europe, added considerable uncertainty as to which energy sources would be dominant in future years. Third, increasing nationalism in the important oil-producing areas of the Middle East and North Africa made them increasingly less dependable as sources of oil for the major companies.

These circumstances endangered the major international oil companies by threatening to deprive them of their monopolistic control over low-cost crude oil supplies, and by threatening to make that control less profitable if other energy sources gained a competitive advantage over oil. In response to this situation, the major international oil companies have each adopted strategies which in effect seek to restore their monopolistic position and profitability by gaining control over all energy sources.

Such a trend toward monopolistic control of all energy sources can arise without any explicit conspiracy among the companies, simply by each company seeking to minimize its risk through diversifying into other energy sources. Thus, historically within the United States, the march toward almost total vertical integration in the oil industry, with refiners moving backward into crude oil production and forward into marketing, was not the result of any grand conspiracy but of each company taking protective measures to insure the stability of its profits. But this does not mean there are no elements of conspiracy. Since collusion may further speed up the rate of monopolization, from an economic point of view it would clearly be desirable for the companies. At this point, I do not prejudge its extent, but merely want to stress that while collusion may exist (see Chapter 8), it is not a necessary element of a strategy which leads toward monopolistic control over all energy sources.

Gaining such control would provide several major benefits to the big oil companies, particularly the American ones. First, they could make monopolistic profits, by driving up their prices; moreover, since there is substantial competition among different energy sources, this would also allow them to maximize the monopolistic profits from their oil. Second, since coal can be transformed into oil or gas, it would give the companies even greater marketing flexibility. And third, it could tend to speed up the growth and guarantee the stability of the companies insofar as coal and uranium would be used relatively more than oil in the fastest-growing and yet most stable market, the electrical utilities.

The major American companies have adopted a number of tactics to implement this overall monopolization strategy. The first and most fundamental step has been to gain control over large quantities of non-oil energy reserves, particularly in the United States. This is a process that has been going on throughout the 1960s, quietly but with remarkable speed. Definitive data are not available because in the United States the figures for energy reserves controlled by individual companies and indeed even by whole industries are confidential, and often not available even to the government. Nevertheless, it is apparent that the major oil companies have sought with a vengeance to gain control over other energy sources and have been extremely successful.

Thus, an official of Exxon described for Congress how his company became involved in coal in the early 1960s:

> Our studies of the nation's energy requirements indicated that utilization of all of the nation's energy resources would be needed to meet increasing demands. We concluded that coal mining and the marketing of coal as a utility fuel offered an attractive long-term investment opportunity which draws upon Humble's [the Exxon subsidiary] experience in exploring for minerals and its established management and technical resources. Humble recognized concurrently that coal at some future date could become a suitable raw material to supplement crude oil and natural gas as an economically attractive source of hydrocarbons.

U.S. Senator Aiken has testified that the oil companies "have now virtually gained possession of the larger mines of this country." The significance of this can be seen from the fact that the United States has the non-Communist world's largest reserves of coal, enough, if the coal is turned into natural gas or oil, to meet its energy needs for many generations to come. Interestingly, even before the October 1973 war, Exxon had predicted that by 1980 one-tenth of all U.S. coal production would be used for producing synthetic gas.

With regard to uranium in the United States, a government official reported that the oil companies had the following percentage shares of various facets of the business by the end of the 1960s: production, 14 percent; exploration and drilling, 40 percent; reserves, 45 percent. While this discrepancy between the oil companies' relative share of uranium production and reserves may partly reflect the necessary time lag between discovery and production, it also could be that the oil companies were stockpiling uranium reserves, looking toward much higher uranium prices in a monopoly-dominated industry in the future.

In addition, the major oil companies are moving rapidly toward vertical integration in the nuclear field, a process which would allow them to realize monopolistic profits from their uranium. For example, Exxon is in nuclear fuel fabrication as well as uranium mining, while Gulf has entered not only mining and fuel fabrication but also into building nuclear reactors and reprocessing spent uranium.

Since natural gas frequently is physically associated with crude oil, it is not surprising that the big American oil companies also have huge resources of natural gas. For example, Exxon reportedly has about one-seventh of total U.S. natural gas reserves.

Once the big oil companies acquired vast quantities of non-oil energy sources, they then had a major incentive to try to drive up the price of all energy resources. One of the most important tactics in achieving this has been the creation of an "energy crisis" atmosphere throughout the world, and particularly in the United States. It is extremely significant to note that the industry propaganda campaign for the "energy crisis" did not begin with the historic Teheran and Tripoli agreements of early 1971, in which members of the Organization of Petroleum Exporting Countries (OPEC) negotiated sizable price increases (see Chapter 8). The campaign was further accelerated by those agreements, but it was actually launched in 1968.

Not coincidentally, this was the same year that the U.S. Supreme Court ruled that the Federal Power Commission should not allow increased natural gas prices as urged by the oil industry, which argued that it needed more money because gas reserves were declining sharply. The Court rejected this argument, noting, "There is . . . substantial evidence that additions to reserves have not been unsatisfactorily low, and that recent variations in the ratio of reserves to production are of quite limited significance." The Court also noted that each year new reserves exceeded production. Since then the industry has reported steadily declining gas reserves. As Robert Sherrill, a perceptive journalist who has carefully reconstructed the industry's "fright campaign," points out, since the 1968 Supreme Court ruling there has been a rising crescendo of advertisements, business publication reports, and popular press articles on the ever more dangerous "energy crisis."

Eliminating all controls on natural gas prices in the United States is probably one of the principal goals of the major oil companies today. This is the reason they are quite willing to have other companies import foreign-government-owned liquefied natural gas (LNG) into the United States. While the big companies temporarily lose markets, the longer-run benefits can be great. As James Ridgeway notes:

> Even before the final arrangements were completed, the El Paso–Algerian deal had stirred considerable repercussions. Because the imported LNG will cost much more than gas piped from fields in the southwestern part of the United States, it contributed to the pressures resulting in higher gas prices within the United States. Indeed, it might be considered to have established a threshold level for this sort of gas.

The price for Algerian liquefied natural gas to be delivered to the United States (before the October 1973 war) was about $1.00 to $1.25 per 1,000 cubic feet, or four to five times the price of domestic natural gas. If the big oil companies could raise the price of domestic gas to anywhere near parity with this Algerian price, it would be worth billions of dollars a year to them in additional natural gas profits. Further, since in recent years natural gas has been the cheapest fuel available, raising its price could also raise the ceiling on prices for oil, coal, and uranium, thereby affording the major oil companies additional billions of profits. Hence a few "arms-length transactions" (deals negotiated between independent parties), like that of El Paso Natural Gas with Algeria, offer the perfect pretext for establishing a generally high level of domestic energy prices.

Finally, as we have seen, at least for the medium to long run (i.e., through the year 2000 at least), there are ample physical supplies of energy available. Hence, the real prospect is not lack of supply, but artificial shortages created by monopolistic control of that supply. It is here that the Teheran and Tripoli agreements of 1971 became so useful to the major oil companies. They were portrayed to the public, particularly in the United States, as an unadulterated victory for OPEC power, and a devastating blow to the profitability of the major international oil companies. The rise in prices which followed the agreements was attributed to the OPEC governments, leaving the companies blameless.

In contrast to this publicized picture, there is considerable speculation within the international oil industry that in fact the major oil companies did not resist the OPEC demands with any great fervor. While we will examine this in detail in Chapter 8, it should be noted here that the companies derived at least two major benefits from these 1971 negotiations and agreements. First, it is a fact that while the OPEC government taxes on oil revenues were increased sharply under the agreements, the companies were able to increase their prices considerably more, so that their per-barrel profits rose even more sharply than the taxes. Second, the U.S. Justice Department's removal of antitrust restrictions on the American oil companies attending the Teheran negotiations, to allow them to prepare a common front against OPEC, opened the door for the companies to allocate and control markets for oil and other energy sources.

The strategy and tactics of the international oil companies raise a number of fundamental questions. First, will they succeed in their attempt to monopolize the world's energy resources and maintain the new level of high prices and profits? Second, what will be the impact on this strategy of the increasing participation of the OPEC countries in their oil industries, which appear headed toward complete "nationalization" in the not too distant future? And third, what will be the impact of the companies' strategy on the nations of the world, both developed and underdeveloped? Before we can begin to answer these questions we must first examine the crucial question of the relationship between the major international oil companies and their home governments, to which we turn in the next chapter.

3

The International Oil Companies and Their Home Governments

One of the great myths of recent times is the notion of the "free-floating" multinational corporation which roams the world like the lord of the jungle, uninfluenced by stockholders or states. Even a cursory examination of the historical and current relationships between the largest of the multinationals, the international oil companies, and their home governments is a useful antidote to this nonsensical and dangerous notion.

As we shall see here, the fate of the international oil companies has been and is closely tied up with the power of their home governments, which have continually exerted this power on behalf of the companies. Moreover, one can speak meaningfully of the "home governments" for these international companies because however widespread the scope of their operations, each company has a clearly defined home country in terms of the majority of stock ownership. Thus, Exxon may operate in every country from Australia to Zambia, and even have many subsidiaries legally incorporated outside the United States, but Exxon is a U.S. corporation because most of its stock is owned by U.S. citizens. Shell Oil, on the other hand, is incorporated in and operates almost totally within the United States, but for all its efforts to identify itself in the public mind as a U.S. company, it is in fact a Dutch-British firm because 69 percent of its stock is owned by Royal Dutch Shell, which in turn is majority-owned by Dutch and British interests. And where the majority of stock ownership rests, the government, one can be sure, will be a more or less active supporter of the company.

There are a number of reasons why this is so. First and

most basic, as we discussed earlier, in many of the developed capitalist countries, particularly the United States and Great Britain, and to a lesser extent France and the Netherlands, the big oil companies represent enormous concentrations of economic power. And long before Watergate it was obvious to anyone with common sense that economic power begets political power, in a variety of ways. Second, there has often been a symbiotic relation between the oil companies and their governments. Thus, while the oil companies seek profits all over the world, their governments also want to assure reliable supplies of oil for their military machines and for their industries, as well as for popular consumption. Each of these aims can be furthered by government support for the expansion efforts of the oil companies. And if in the process of expansion the oil companies bring in huge profits from abroad which help the home country's balance of payments, this too gives the government a stake in the companies' operations.

The perception of such a symbiosis is made easier by the two-way flow of personnel which has taken place between the oil industry and government agencies dealing with oil and foreign policy. The importance of this flow was enunciated openly by Exxon's treasurer after World War II:

> As the largest producer, the largest source of capital, and the biggest contributor to the global mechanism, we [the United States] must set the pace and assume the responsibility of the majority stockholder in this corporation known as the world . . . American private enterprise . . . may strike out and save its own position all over the world, or sit by and witness its own funeral. . . . As our country has begun to evolve its overall postwar foreign policy, private enterprise must begin to evolve its foreign and domestic policy, starting with the most important contribution it can make—"men in government."

The two-way flow between the oil companies and the government has been described by Robert Engler as follows:

> The permeability of oil has extended throughout the peacetime machinery of the federal bureaucracy. . . . The number of public

officials with oil backgrounds or relations could begin with former Secretary of State John Foster Dulles. Until 1949 he was the senior member of Sullivan and Cromwell, the major law firm for the Jersey Standard empire.... Herbert Hoover, Jr., a petroleum engineer and director of Union Oil whose major associations have been with oil, was the State Department's representative in the secret Iranian negotiations. He later became Undersecretary of State and was involved in questions of Middle Eastern policy and represented his Department in many of the top-level Suez arrangements. Winthrop W. Aldrich, head of the Chase Bank, which has long been tied to the Rockefeller and related oil interests, was sent to London in 1953 as United States Ambassador. In addition to the general sympathy within the State Department for oil positions, there has also been an interchange of personnel with the industry. For example, William A. Eddy, former educator, OSS chief in North Africa and the first full-time United States resident Minister to Saudi Arabia, became a consultant to the Arabian American Oil Company, handling governmental and public relations. Harold B. Minor, once ambassador to Lebanon, became assistant to James Terry Duce, Aramco's vice president for government relations. Brigadier General Patrick J. Hurley has been an attorney for Sinclair and a special envoy to the Middle East. Henry F. Holland, former Assistant Secretary of State for Inter-American Affairs, has represented oil groups involved in Latin America.

Walter J. Levy, the first chief of the European Cooperation Administration's Petroleum Branch, was a petroleum consultant whose clients included Esso, Caltex, and Shell. He had also been on Averell Harriman's staff when the latter was sent by President Truman to seek a solution to the Iranian dispute....

Robert B. Anderson, formerly Secretary of the Navy, Deputy Secretary of Defense, and appointed in 1957 Secretary of the Treasury to succeed George M. Humphrey, was a Texan who had been active in oil production.... Anderson had been manager of the $300 million W. T. Waggoner estate with its extensive oil operations, a member of the National Petroleum Council and a director of the American Petroleum Institute....

Wherever there are consultants called in on national policy, the position and power of oil within the business community are sure to be recognized.

The most recent example of the close links between the oil

companies and the U.S. government came to light with the formation of the Federal Energy Office (FEO) after the October War. In a story entitled "Links of Some Federal Energy Experts to Industry Raise Questions of Conflict," the *Wall Street Journal* noted, among a number of cases, that of an Exxon senior adviser on government affairs, specializing in the Middle and Far East, who left Exxon in January 1974 to join the FEO. His new job is to serve as acting assistant administrator for foreign affairs, in which post he will deal with both foreign governments and oil companies like Exxon. Moreover, before leaving Exxon this executive received a lump sum payment of tens of thousands of dollars from the company. A spokesman "says Exxon and a few other large companies make such payments so executives won't be discouraged from taking government positions."

Finally, of course, there is in some countries a direct relationship between the oil companies and their government. Thus, the British government owns half of British Petroleum, one of the Big Seven, and the French government owns 35 percent of Compagnie Française des Petroles (CFP), sometimes considered one of the Big Eight. In Italy, Ente Nazionali Idrocarburi (ENI) is a state oil corporation, while Deminex and Veba in Germany and the Japanese Petroleum Development Corporation are subsidized or owned by their respective governments.

The fact is that the Big Seven's rise to dominance is primarily a function of the growth of U.S. and British power in the twentieth century, and the recent dominance of U.S. over British companies itself reflects the relative weakening of British power as a result of World War II. While the United States had a great natural advantage over the rest of the world because of its large oil production prior to World War II, the big growth in world production since then, as we have seen, has come outside the United States, particularly in the Middle East. Thus it seems quite likely that had the Axis won World War II, the prolific Middle East oil fields, and hence the world oil industry, would be dominated by German, Japanese, and Italian interests.

The role of state power in buttressing the international oil companies can be traced virtually to the origins of the industry. The first challenger to Standard Oil was the "Royal Dutch Company for the Working of Petroleum Wells in the Netherlands Indies" (the forerunner of Royal Dutch Shell). It was organized with Crown support in 1890 to exploit the oil of that Dutch colony. In order to protect the fledgling Royal Dutch, the Dutch government blocked Standard from entering the colony and sold the oil concession to Royal Dutch for $45,000, although it was worth almost $25 million. When Standard tried to gain control of Royal Dutch by buying up shares in the open market, the company quickly issued 1,500 shares which had complete voting control and could be held only by Dutch subjects, thereby insuring its continuation as a Dutch company. (Interestingly, the head of Royal Dutch during its period of continuous battles with Standard Oil had a motto which clearly reflected a nation-state outlook: "Either war everywhere, or peace everywhere.")

Again, Shell Transport and Trading, a British firm which was one of the Big Three before it merged with Royal Dutch, got its big breakthrough in 1892 when it was allowed to use the Suez Canal to transport Russian oil in tankers (rather than in barrels) to the Far East. Despite fears of explosion, the permission was granted because Shell was a British company and the Canal was controlled by Great Britain.

British Petroleum, which got its start in Iran in the early 1900s as the Anglo-Persian Oil Company, owed its very life to British state power. British gunboats and troops protected the key British concession, the D'Arcy concession of 1901, from its archrival in Persia, Russia. When, after discovery of the world's largest oil field in 1907, the concessionaires were ready to quit in 1909 because of the mammoth problems of developing and marketing the oil, a British officer wrote to the British official in charge of the Empire around the Persian Gulf as follows: "It amazes me that the director . . . should be in a position to risk the complete loss of a concession . . . without consulting with the F.O. [Foreign Office] and without telling you or the Minister or the Government of India.

They have all the vices of absentee landlords. Cannot Government be moved to prevent these fainthearted merchants, masquerading in top hats as pioneers of Empire, from losing what may be a great asset?"

As a result, the British government got Burmah Oil, a Scottish company, to launch Anglo-Iranian with an investment capital of £2 million. Then, in 1914, Winston Churchill, First Lord of the Admiralty, in order to assure a wholly British-controlled source of oil for the navy, induced the British government to invest another £2 million in Anglo-Iranian and take over majority control of the company (although operation of the company was left in nongovernment hands). This was in line with the British government's explicitly stated aims: "For many years the policy of the Foreign Office, of the Admiralty, and of the Government of India has been to conserve independent English interests in the region of Persia, and, above all, to prevent this region falling under the control of the Shell or any other foreign or cosmopolitan company."

The American government, on the other hand, fought a continuing battle on behalf of U.S. oil companies to get them a foothold in Iran. For example, in 1920 the U.S. Secretary of State cabled the British Foreign Office the State Department's position that "the monopolization of the production of an essential raw material, such as petroleum, by means of exclusive concessions or other arrangements" was contrary to the Open Door policy of the United States. When in 1921 Standard Oil of New Jersey (now Exxon) applied for a concession in Iran, the Foreign Office protested to the State Department. Clearly, only British government power could create a situation where the former U.S. Ambassador to Iran could comment that it was "an interesting and striking fact that the four vigorous efforts of friendly American oil companies in 1921, 1923, 1936, and 1944 to obtain oil concessions in Persia were all, in turn, unavailing despite the undoubted fact that no great power in recent times has ever enjoyed the general and spontaneous good will felt for Americans by Persians of all classes." Only when British power had been drastically weakened by World War II could British Pe-

troleum's monopoly in Iran be broken, as will be described in a
later chapter.

A similar struggle for oil resources took place in Mesopotamia (now Iraq), which was controlled until the end of
World War I by Turkey and was one of the prizes of that
bloodbath. The struggle was particularly sharp because after
the war Standard Oil of New Jersey, which had traditionally
emphasized marketing abroad, now began avidly to seek overseas production of crude oil as well. The British government,
disposing of the main military power on the scene, blocked
Standard's and Mobil's exploration efforts while giving full
support to those of Royal Dutch Shell and British Petroleum.
(This caused Standard's president to deliver a gem of business
patriotism: "British domination would be a greater menace
to [Standard of] New Jersey's business than a German victory would have been.") The British move led to a hot exchange of notes and pressures between the two governments,
with the State Department telling the U.S. companies, according to a Gulf Oil official, "to go out and get it," and a
British Foreign Office official claiming that "Washington officials began to think, talk, and write like Standard Oil officials."

After years of complex negotiations, Iraq's oil was finally
conceded, as part of the 1928 "as is" agreement, to a joint
company in which Standard (Exxon) and Mobil together got
one-fourth, with the remaining quarters given to British Petroleum, Royal Dutch Shell, and CFP (with 5 percent of each
company's share going to the entrepreneur Gulbenkian).
Thus, thanks to government, U.S. oil interests got their first
foothold in the Middle East.

Around the same time, an even more significant breakthrough was made by the State Department for U.S. oil companies. On the island of Bahrein, off the Arabian shore, Standard of California obtained Gulf Oil's concession, which the
latter could not use because of the "as is" agreement. Standard was then blocked by the British Colonial Office, which
declared that only a British-managed company could have the
concession. The State Department took up the battle for the

Open Door policy, and a compromise was finally reached by which a Canadian subsidiary of Standard of California took over the concession, while the British government was given certain administrative rights.

While Bahrein did not become an important oil producer, it was a springboard for American entry into the two golden concessions of the Middle East, Saudi Arabia and Kuwait. Standard of California, which found oil in Bahrein in 1932, in 1933 topped the Iraq Petroleum Company's offer of $45,000 to King Ibn Saud with its own bid of about $250,000 in gold sovereigns. For this amount Standard of California got a sixty-year concession for all of Saudi Arabia's oil—a far better bargain than even the fabled deal giving the American Indians $24 for Manhattan island. By 1936 Saudi Arabia looked so promising that Standard of California, which had no outlets for vast quantities of crude oil, sold a half-share of its Bahrein and Saudi properties to Texaco for $3 million down and $18 million out of future earnings— another great bargain, this time for Texaco. The partnership between the two companies is known as Caltex.

British and American rivalry heated up in Saudi Arabia during World War II. The war brought Saudi production to a virtual standstill, but Ibn Saud still needed oil revenues. Caltex tried to get the U.S. government to give him the money, and this was arranged at first by having the British pay him out of U.S. lendlease loans. However, because of U.S. fears of the British, who by then had sent a team of geologists into the country along with a military expedition, in 1943 the lendlease aid was given directly by the United States. Petroleum Administrator Harold Ickes wanted the U.S. government to set up a corporation to take over Caltex's interest, but the oil companies had this defeated in Congress.

On the other hand, the U.S. government, in its eagerness to advance the oil companies' interests and block the British, provided huge amounts of very scarce steel after World War II for Aramco to build a thousand-mile pipeline to the Mediterranean. In addition, after the companies were forced to sharply increase their tax payments to Ibn Saud in 1950, under a new 50-50 profit-sharing agreement, the U.S. government allowed the companies to deduct the payments from their tax obligations at home. The increase therefore cost the companies virtually nothing; the average American taxpayer footed the bill.

If Saudi Arabia was an unequivocal victory for the United States, in Kuwait compromises again had to be made with British power. An independent entrepreneur had gained the concession in all of Kuwait in the 1920s, and after British Petroleum refused to buy it from him he offered it to Gulf which, having been blocked from Bahrein, readily accepted. Before Gulf could nail down a deal, however, the British government, which had sovereignty in Kuwait, intervened to block it. By this time British Petroleum had a new interest in Kuwait, and a struggle developed between the British and U.S. governments over who was to get the concession. Interestingly, the American effort was spearheaded by Andrew Mellon, U.S. ambassador to Britain, whose family just happened to control Gulf Oil. In any event, the long tug of war was ended in 1933 when Gulf and British Petroleum settled on a 50–50 joint venture, the Kuwait Oil Company, to take over the concession.

Lest the reader mistakenly get the impression that the Middle East was the only arena of struggle, and the United States and Great Britain the only parties, we can cite the case of the Netherlands Indies. Here, after World War I, the Dutch government was planning to turn Germany's concessions over to Royal Dutch Shell, freezing out American companies. A U.S. State Department memo described the American reaction: "With a view to exerting further pressures on the Netherlands Government the American Government took steps to block the issuance of further concessions to the Royal Dutch Shell in public lands in the United States. The final result was that the New Jersey company [Exxon] was given additional producing concessions in the Indies which turned out to be some of the richest in the islands." Another example: Since France's main international oil company, CFP, was set up in 1924 to hold France's share of former German holdings in the Ottoman Empire, it is self-evident that this major company too derived its very birth from its home government's power.

World War II, of course, decisively altered the relative power of various governments, and this shift was mirrored in the relative position of each country's oil companies. The simultaneous strengthening of U.S. power relative to that of all other developed capitalist countries, combined with the upsurge of Soviet power and of nationalist and anti-imperialist sentiments in the Third World, made it both inevitable and necessary for the U.S. government to push American oil companies to the fore. As we noted earlier, one display of this power was the rapid penetration of world energy markets by oil, particularly in Europe where the international oil companies were aided by U.S. government loans under the Marshall Plan.

Another area where U.S. power was exerted was in the opening up of the former European colonies to U.S. oil interests. The drive began with wartime lendlease programs, under which the European powers had to accept the principle of the United States having "free and equal" access to all raw materials in their colonies. U.S. pressure was further exerted through the leverage of the Marshall Plan, under which each recipient country had to guarantee American capital equal opportunity with its own citizens to invest in the development of colonial raw materials. As a result of these kinds of pressures, American oil companies got footholds or enlarged their positions in many countries which had been preserves of European capital. Thus, British domination of India's oil market and Dutch domination of Indonesia's were broken, while American companies got oil concessions in such places as French Tunisia, Portuguese Mozambique and Angola, and British-dominated Ethiopia.

Washington's eagerness to promote American oil interests, even in countries like Mexico and Bolivia which had nationalized their oil industries, was unbounded. At the end of 1950 the U.S. Ambassador to Bolivia reported triumphantly, if somewhat prematurely, that: "Since my arrival here, I have worked diligently on the project of throwing Bolivia's petro-

leum industry wide open to American private enterprise. . . .
The whole land is now wide open for free American enterprise. Bolivia is, therefore, the world's first country to de-
nationalize. . . ."

However, the most important arena of power struggle in
the Third World was still the Middle East. The U.S. aim was
to muscle Britain out of as much of its oil holdings as possible, yet not make it so weak or uninterested in this crucial
area as to create a vacuum which might be filled by the
Soviets or by radical nationalism. Owing to the vast discovered reserves of Saudi Arabia and Kuwait, the U.S. share of
total Middle Eastern reserves had jumped from only 10 percent in 1940 to one-third in 1945, and one-half in 1950.

Yet the United States still coveted Great Britain's 100 percent control of Iranian oil, which in 1950 made up over
one-fifth of all Middle Eastern reserves. The opportunity to
do something came in 1951 when an international crisis was
precipitated by the Iranian government's nationalization of
its oil industry. Because this was such a signal event in the
history of the international oil industry, and prototypical of
the role of state power in assisting the international oil companies, it is worth examining in some detail.

The long-run factors leading up to the nationalization consisted essentially of a widespread belief among Iranians that
they got little benefit from the oil industry, and that without
control of it they could not rule their own destiny. Related
to this were longstanding grievances over the way Iran, historically a political football between Great Britain and Russia,
had been forced or "tricked" into unfair oil agreements.

The immediate backdrop to the 1951 crisis was the government's drive to gain greater oil revenues for financing its
Seven-Year Plan, inaugurated in 1949. In addition, by 1949
Venezuela and Saudi Arabia had already won 50–50 profit-
sharing agreements with the international oil companies,
while Iran was still receiving a much smaller share. Since
1947 Anglo-Iranian had accepted that some changes would
have to be made in the Iranian concession agreement, but it

dragged its heels until it was too late to stave off the nationalization by the Mossadegh government.

The British government found itself in a quandary as to how to deal with the situation. On the one hand, the use of naked force was ruled out by the changing world power structure and particularly by the danger of Soviet intervention. On the other hand, Great Britain was unwilling to simply accept Iran's offers for compensation, both because it might set a precedent for nationalization in other countries, and also because any reasonable settlement for the physical properties of the company would be inadequate to compensate the British economy. Thus, while the company had total assets of about $1.5 billion, part of which lay outside Iran, it not only made large profits but also allowed the dollar-short British to pay in sterling for $300 million of oil per year.

Great Britain thus first took its case both to the International Court, which held that it did not have any jurisdiction in the case, and then to the Security Council, which also took no effective action. While pursuing these channels of appeal, Great Britain initially placed its greatest hope on the cumulative impact of various economic retaliations, particularly the boycott it organized on the export of Iranian oil.

The boycott was extremely effective in cutting off virtually all of Iran's oil income. At the same time, neither Anglo-Iranian nor Great Britain suffered the losses that might have been feared, owing to the company's ability to draw upon its fabulous oil reserves in Kuwait. On the other hand, while the oil boycott caused financial trouble for Iran, it was not directly decisive in achieving its aim of overthrowing the Mossadegh government and/or restoring the nationalized properties to the company. The basic reason was that the oil sector was a relatively autonomous island in the Iranian economy, affecting only a small number of Iranians; moreover, owing to the government's small share of oil profits, oil revenues accounted for only about 12 percent of total government revnues. However, in the long run the boycott helped to undermine the government insofar as it forced it to take measures which contributed to internal political discontent, particularly on the part of the Iranian upper class:

> Needing money desperately, on August 21, 1952, the Premier decreed the creation of a commission to collect the immense arrears of unpaid taxes of the rich for the last ten years. The commissions were empowered to throw the wealthy Iranians, so called, "thousand families" in jail and confiscate their property if they did not pay. An additional blow to them was Mossadeq's decree that they cut feudal dues received from sharecroppers and return 10 percent of the profits derived from the land they worked. . . .
>
> About this time [early 1953] Mossadeq was disputing bitterly with the monarchy over the question of the control of the army. In an effort to minimize the power of the king and to ease the country's financial woes, Dr. Mossadeq decided to cut into the Shah's $720,000-a-year government allotment and his $2,000,000-a-year income from other sources. The tension between the Shah and Mossadeq was intensified. . . .
>
> A chain of ominous events followed one after another. . . .

The decisive cause of the overthrow of the Mossadegh government in 1953 was the actions of the U.S. government. David Wise and Thomas Ross, while noting that "the British and American governments had together decided to mount an operation to overthrow Mossadegh," describe graphically the crucial role of the U.S. Central Intelligence Agency:

> There is no doubt at all that the CIA organized and directed the 1953 coup that overthrew Premier Mohammed Mossadegh and kept Shah Mohammed Reza Pahlevi on his throne. But few Americans know that the coup that toppled the government of Iran was led by a CIA agent who was the grandson of President Theodore Roosevelt.
>
> Kermit "Kim" Roosevelt, also a seventh cousin of President Franklin D. Roosevelt, is still known as "Mr. Iran" around the CIA for his spectacular operation in Teheran more than a decade ago. . . .
>
> One legend that grew up inside the CIA had it that Roosevelt, in the grand Rough Rider tradition, led the revolt against the weeping Mossadegh with a gun at the head of an Iranian tank commander as the column rolled into Teheran.

A CIA man familiar with the Iran story characterized this as "a bit romantic" but said: "Kim did run the operation from a basement in Teheran—not from our embassy." He added admiringly: "It was a real James Bond operation."

One of the significant aspects of the Iranian boycott was the demonstrated solidarity of the international oil companies with their British "competitor." The rewards to the oil companies for their solidarity were substantial. Following the overthrow of the Iranian government, rather than the oil properties being restored to Anglo-Iranian, a new corporation was given control of the properties. Only 40 percent of the shares of this corporation went to British Petroleum, with 14 percent going to Royal Dutch Shell, 6 percent to CFP, and 8 percent to each of the five U.S. majors, who in turn gave one-eighth of their share to eleven other U.S. companies. As a footnote to the whole Iranian affair Kim Roosevelt "later left the CIA and joined the Gulf Oil Corporation as 'government relations' director in its Washington Office. Gulf named him a vice-president in 1960."

The crucial Iranian events, along with the Suez Canal fiasco of 1956, sealed the end of overt U.S.-British rivalry over Middle Eastern oil affairs. Britain had to accept its junior partner role in the area, and after that worked jointly with the United States when action was necessary.

A good example of this joint action came in the summer of 1958 when a military group overthrew the feudal dictatorship in Iraq, and the entire Middle East was in ferment. American marines and a naval armada were dispatched to Lebanon, where a civil war threatened, and British paratroopers landed in Jordan. Despite the American government's claim that it sent the marines to protect Lebanon from "foreign agents," there is no question but that oil interests were the precipitating force. Sir Anthony Eden wrote afterward: "Since the United Nations observers were already on the spot and proclaiming that the motives for Anglo-American intervention did not exist, it was rather more heinous."

Indeed, according to the *New York Herald Tribune*, at first the American government gave "strong consideration" to "military intervention to undo the coup in Iraq"; the State Department advised the U.S. ambassador to Lebanon that "marines, starting to land in Lebanon, might be used to aid loyal Iraqi troops to counterattack." Unfortunately for the United States, no Iraqis could be found to act as tools for restoring a universally detested regime.

Nevertheless, the threat to the new government was clear. The *New York Times* reported the decision of conferences between President Eisenhower, John Foster Dulles, and Foreign Secretary Lloyd of Britain: "Intervention will not be extended to Iraq as long as the revolutionary government in Iraq respects Western oil interests." As Robert Engler comments, "This gunboat diplomacy was clearly in line with the State Department's commitment to pipelines and profits." What other position could be expected from corporate lawyer Dulles, who during the 1956 Suez crisis declared to a secret meeting of top oil company executives and government officials that as regards U.S. oil holdings in the Middle East *"nationalization of this kind of an asset* impressed with international interest was far beyond compensation of shareholders alone and *should call for international intervention."* Nevertheless, the continuity of American policy can be seen from the fact that the new Iraqi government of Colonel Kassem was overthrown in early 1963 by another coup. This coup followed right on the heels of Kassem's announcement of the formation of a state oil company to exploit oil lands seized from the companies in 1961; it came four days after Kassem revealed an American note threatening Iraq with sanctions unless he changed his position. He did not, and the Paris weekly *L'Express* stated flatly that "the Iraqi coup was inspired by the CIA."

Just when the ancient Anglo-American rivalry was being buried under the weight of U.S. power, faint stirrings of competition were beginning in several important developed countries, particularly Italy, France, and Japan, as well as in the Soviet Union. This competition in turn was increasingly associated with a renewed if limited drive for more independence on the part of the oil-exporting underdeveloped countries. A clear understanding of the complex position of these producing countries is vital to comprehending the possible future of the energy crisis; we therefore turn in the next chapter to an analysis of how this position evolved.

4
The Oil-Exporting Countries

Since the October War the people of the capitalist world have suddenly found the fate of their economies as well as their own personal comfort apparently hinging on a small handful of oil-exporting underdeveloped countries. With a sense of shock soon intensified by the West's deeply embedded racism, the people reacted with frustration and anger at these hitherto-unknown dark-skinned "desert barbarians" who now seemed to have so much power. The ignorance of people in the West about the oil-exporting underdeveloped countries reflects the veil of secrecy which the Western ruling powers have drawn about their imperialist activities, for these governments are no strangers to any facet of economic activity in the oil-exporting countries. The fact is, as we shall show, that for most of the twentieth century the fate of the oil-exporting underdeveloped countries has been almost totally shaped by the forces of Western imperialism—specifically, the international oil companies and their home governments.

Thus, historically, wherever there was a promise of oil in the Third World, the international oil companies were there to seek concessions. Usually the companies got a foothold in the countries by dealing with unrepresentative and illegitimate rulers. The original oil concessions were often obtained by illegal means. Moreover, they were granted at terms so unfavorable to the interests of the country as to be almost fraudulent.

The history of oil concessions as chronicled by relatively objective Western scholars reveals clearly that this was true in nearly all such concessions. Taking the Middle East as a key example, George Stocking, an authority on oil in the area, has some revealing observations as to the representativeness of the rulers who granted oil concessions.

With regard to Iran, Stocking notes that the shahs traditionally "ruled with iron, frequently bloody, but uneasy hands. Theirs was an absolute despotism in which the people and the land were regarded as the property of the shah to do with as he chose." He further notes that the crucial Iranian oil concession of 1901 was granted to the British entrepreneur D'Arcy by "an irresponsible and despotic ruler, unsophisticated in business affairs, [who] was more interested in meeting his and the royal family's needs than in protecting the patrimony of his people." Even the Persian government, which in 1933 negotiated the existing extension of this 1901 concession, "pointed out that the original concession had been granted by a government that had no constitutional basis."

As for Iraq, 1925 was the year when the country was nominally granted "independence" by Great Britain, and the existing concession agreement was signed. As Stocking notes,

> The king and his cabinet ratified the new concession on March 14, 1925, one week before King Faisal promulgated Iraq's organic law and approximately seven months before the Iraqi parliament ratified the treaty defining Iraq's obligations under the mandate principle; in short before an autonomous Iraq government had anything to say about it.

Little needs to be said about the clear illegitimacy not only of the sheiks and kings who granted oil concessions in the rest of the Middle East, but also of the military dictators of Latin America and elsewhere who did the same. For example, General Juan Gómez, who granted most of Venezuela's oil concessions, was a dictator who seized power from the incumbent despotic president in 1908 and held it through 1935, according to O'Connor, "protected by perhaps the most savage set of cut-throats to be found in all the sad history of Latin America."

Aside from the question of governmental legitimacy, it is also transparent that many major oil concessions were based on bribery, fraud, and/or duress. In the case of Iran, for example, Stocking notes:

> While the Persian shahs of the nineteenth century customarily were neither accountable to nor held in restraint by any coherent social group, they were not equally free from foreign influence. Foreign powers vied with each other in trying to obtain a preferred economic and geographic position in Persia and to utilize Persian military forces to strengthen their imperial dominions. . . . Russia and Great Britain settled their rivalry in Persia by signing a treaty, the Anglo-Russian Convention of 1907, which, while piously recognizing Persian integrity and independence, divided the country into spheres of Russian and British influence.

In 1909 the Shah signed an agreement granting pipeline right of way and selling Abadan Island to the British-owned Anglo-Persian Oil Company for a mere pittance. More important, as Stocking notes, "the terms negotiated by a representative of the British government plainly disregarded the letter and spirit of the D'Arcy concession." In more modern times, the Iranian oil industry, denationalized after the 1953 overthrow of Mossadegh, clearly was the illegitimate offspring of Western economic and military power.

In Iraq, Great Britain was the thinly veiled power behind the throne which granted the key concession in 1925. In the words of the British High Commissioner to Iraq: "It was their [Great Britain's] intention not that the proposed treaty [granting Iraq independence] should replace the Mandate, but rather that the Mandate should be defined and implemented in the form of a treaty." As Stocking points out,

> Before the king ratified the treaty, protests against it became so vigorous as to occasion the resignation of the king's cabinet, the assumption by the British high commissioner of dictatorial powers, the deportation of the treaty's most vocal opponents, and the appointment of a new cabinet whose "sole purpose" . . . was to ratify the treaty.

In the case of Venezuela, according to Edwin Lieuwen, a leading expert on the history of the Venezuelan oil industry,

Fraud and deception, chicanery and double-dealing, were the rules of the game. . . . Three big American companies [Standard of New Jersey, Standard of Indiana, and Gulf] obtained their lucrative leases in the corrupt-concessions era of the Gómez regime. Shell rested on its plentiful grants received earlier.

In addition to these violations of legality, many oil concessions were granted on ludicrously unfavorable terms to the underdeveloped country. For example, the Persian concession of 1901 covered 480,000 square miles, or four-fifths of the country. The oil companies were also granted virtually a free hand, including exemption from all taxes. In exchange for this enormous benefit, the Shah and his friends personally got £20,000 sterling in cash and £50,000 worth of stock in the new oil company formed to exploit the concession.

The Persian government was also to get 16 percent of the annual net profits. However, these net profits were computed by the company and were manipulated to keep tax payments at a minimum. The concession was revised in 1933 to run until 1993 with the basic payment being changed essentially to a fixed fee per ton of crude oil produced. What this meant in practice was that, in 1947 for example, the Persian government received about $1.70 per ton at a time when the value of the oil produced was over ten times that figure.

Similar patterns were set for the rest of the Middle East. For instance, in Iraq the key concession granted in 1925 covering most of the country was for seventy-five years. Payment was basically four shillings per ton of oil produced. For this enormous giveaway, King Faisal of Iraq got a "little present" of £40,000 sterling upon signing the concession agreement. In Venezuela, an early 1913 concession covering 4,000 square miles in the productive Lake Maracaibo district was awarded to Shell with no additional taxes ever to be levied beyond a flat $380,000 per year! All of these cases bear out the conclusion of Stocking that "never in modern times have governments granted so much to so few for so long"—and, I might add, for so little.

It is instructive to compare the policy of the U.S. government, which can hardly be accused of being a foe of the oil companies, in leasing U.S. public lands for oil exploration. As Stocking has noted:

> The prodigal nature of the Middle East concessions can be more readily appreciated by comparing them with the United States' policy in granting oil leases on public lands. Our federal policy has been designed to prevent monopoly exploitation of oil lands. Federal laws limit the area covered by exploration permits (not more than 246,080 acres in any one state except Alaska, where the limit is 300,000 acres). They limit the area a prospector may lease after the discovery of oil to one-fourth of the lands embraced in the prospecting permit with a minimum of 160 acres. They limit the terms of a lease to twenty years. They require that three-fourths of the acreage covered by an exploration permit, to which the original lessee has a preferential right, be leased by competitive bidding with a minimum royalty of 12½ percent. They require that lands located in a known geological structure of a producing field be leased by competitive bidding in units of not more than 640 acres. They limit the duration of a lease resulting from competitive bidding to five years or as long as oil or gas is produced in paying quantities.

Once the international oil companies, with the aid of their home governments, got their concessions, they then exploited the countries ruthlessly. Thus, the profits made by the international oil companies from the oil-exporting countries have been so enormous that in most cases the companies have long ago taken out far more money than they ever put into the countries. Therefore, the so-called "foreign investment" in these countries typically is, in fact, the plowed-back investment of the country's own resources.

Evidence for this is indisputable. On an overall basis, two oil experts have calculated that in the Middle East from 1900 to 1960 the international oil companies reinvested only about $1.7 billion of their net income in the expansion of the region's oil industry, and shipped out $14.6 billion as profits.

In the case of Iran, another oil expert estimates that during the 1955–64 period the Western oil companies earned a compound interest rate of profit on "their investment" of almost

70 percent per year. With regard to Iraq, the same expert estimates a rate of profit for the 1925–63 period of 13 to 15 percent per year, on a compounded basis. While this rate of profit may not sound too high, what is too often overlooked is that even if the Western consortium in Iraq earned "only" 15 percent per year on a compounded basis, while its cumulative investment over forty years was less than $1 billion, it had already got back over $2.5 billion. Moreover, since the Iraqi concessions were not to expire until about the end of the century, if profitability continued at the level of the early 1960s, the companies would net an additional $7 billion. Finally, the underlying data reveal that the peak amount of investment made by the companies prior to receiving profits sufficient to finance all future investment was less than *$50 million!* Thus in Iraq the foreign investor's seed capital of less than $50 million would ultimately return him about $10 billion, or 200 times more; this is the stark reality underlying the bland notion that a foreign investor has earned "only" 15 percent per year over a long period of time.

As if this were not bad enough, we must also take into account the economic, political, social, and psychological damages done to the underdeveloped countries as a result of their historic domination by the international oil companies in conjunction with their home governments. One aspect, to be discussed more fully later, is the corruption of local leaders by the foreign powers. Another is the direct foreign governmental intervention in the affairs of the underdeveloped countries. Aside from the outright colonial situations and the cases already noted, I would mention the following outstanding examples: the landing of British and American troops in the Middle East in 1958 when a revolution in Iraq threatened Western oil interests; U.S. support to the Sumatran rebels in oil-rich Indonesia in 1960; and the various attempts to overthrow Castro, among whose gravest sins in U.S. eyes was his nationalization of the foreign-owned oil refineries.

Lesser known perhaps, in terms of the role of oil at least, is the case of the U.S. overthrow of the Arbenz government in Guatemala in 1954. Not only had this government dared to expropriate 234,000 acres of uncultivated land owned by United Fruit Company, the dominant factor in the economy, but also the previous regime had alienated U.S. oil companies with a law providing that only companies with majority indigenous ownership could explore for oil within the country, and no crude oil could be exported; as a result, U.S. oil companies withdrew from Guatemalan exploration.

In any event, the new government installed by the CIA proved most accommodating to major U.S. corporate interests. Almost immediately all the land that the Arbenz government had taken from United Fruit was restored to it. Within six months the country's oil laws were changed to allow crude oil exports, and half the area of Guatemala was quickly thrown open for oil concessions. So little attempt was made to hide the real power behind the new government that these oil laws were drafted in English and presented for approval to the Guatemalan Congress in that form; only the request of one deputy who still retained a bit of dignity led to their being translated into Spanish!

Violent interventions are simply the dramatic tip of the iceberg. Beneath the surface there is a process of continuing intervention by the developed countries, in support of their investors, in the political life of underdeveloped countries. Moreover, key international organizations such as the World Bank and the International Monetary Fund, both controlled by the developed countries, frequently play major roles in pressuring underdeveloped countries on behalf of foreign investment. Since in the case of the world's dominant foreign investor, the United States, 40 percent of all of its investment in the underdeveloped countries is in oil, and more than 60 percent of all of its profits from the underdeveloped countries derives from oil, it is clear which industry is particularly served by the ongoing efforts of these international organizations.

In addition, foreign investment causes damage to the host country by serving in a number of ways to strengthen the forces within an underdeveloped country which are opposed to the social and economic changes necessary for genuine development. First of all, it creates groups which economically are directly dependent upon the foreign companies, e.g., the indigenous employees and local suppliers. Second, part of the governmental apparatus itself often becomes corrupted in favor of the status quo. In particular, the underdeveloped countries' armed forces can easily become unduly dependent upon the developed countries' military advisers and expensive military equipment—a type of "aid" which frequently follows in the wake of major foreign investment.

Finally, the groups which have traditionally ruled many of the underdeveloped countries—the local agricultural and commercial oligarchies—derive renewed strength to block change from the support of the new groups dependent on foreign investment. A natural harmony of interests springs up because it is clear to all that if local property rights are undermined in a genuine agrarian reform, then the property rights of foreign investors in mineral lands and industrial properties may also be endangered. As an example, it is no mere coincidence that the promulgation of an agrarian reform law shortly after the new Cuban government came to power, in the words of the *Wall Street Journal,* "crystallized American opposition to Prime Minister Castro."

No discussion of the harm caused an underdeveloped country by foreign investments would be complete without some consideration of the less obvious psychological and cultural damage which it so often causes—damage which not only is serious in itself but also hinders the development of the economy. It is important to have an understanding of the mechanisms through which these cultural and psychological problems arise. At the risk of some oversimplification, the situation may be viewed as follows. Objectively, foreign investment involves a two-way traffic between the developed and the underdeveloped country. Flowing into the underdeveloped country, initially at least, are capital plus technological know-how embodied in technicians from the developed world. Later, from the underdeveloped country's tax share of the foreign investment profits, various kinds of consumer comforts and luxuries also flow in. The outflow primarily consists of the raw materials of the underdeveloped country, along with much of the profits made on their sale. Moreover, the outflow is normally one that proceeds from the "backward" hinterland of the underdeveloped country to its relatively modern seaboard metropolises and then to the developed world. One result of this pattern of investment is that often the underdeveloped country's economy tends to be further polarized, with the gap in living standards and culture between the rural areas and the seaboard metropolises being sharply increased.

The subjective corollary of this is that attitudes tend to develop within the underdeveloped country that all good things—namely, the advanced foreign technology and the luxury goods—come from overseas. Moreover, since the fruits of foreign investment come to rest in the big cities of the underdeveloped country, a "pecking order" develops involving rural envy of the big cities and big city envy of the developed countries.

Among the more serious tangible consequences is the tendency for able people in the underdeveloped countries to emigrate first to the cities of their own country and then out to more developed countries. Partly, of course, emigration is due to the objective fact that opportunities for employment are generally correlated with the level of economic development. But, in my view at least, some emigration is caused by the subtle, and often subconscious, attitude that what is rural is bad and what is urban is better and what is foreign is the best. Thus, even in countries where genuine economic opportunities do open up for developing the country, many competent people hesitate to undertake the task. And, of course, where economic opportunities are not readily available at home, the mere presence of the foreign investor, which is usually a multinational corporation, often leads to the much-publicized "brain drain" from underdeveloped to developed countries.

Equally insidious is what is called the "psychology of dependency," a phenomenon observable among even well-trained and skillful technicians in the underdeveloped countries who feel inadequate to cope with advanced technological processes without supervision and assistance from foreign techcians. Sometimes, of course, these feelings have an objective basis in that they have been created by the international corporations' deliberate policy of keeping key elements of technical know-how beyond the reach of the underdeveloped country as a deterrent to nationalization. In the words of Sheik Abdullah Tariki, speaking at the Second Arab Petroleum Congress in 1960, the oil companies "treat us like children." However, from the vantage point of the international corporation, it is only a prudent precaution to leave key blueprints in New York, London, or Paris, and fly in foreign technicians and components when required.

Even where foreign investors do not deliberately set out to create conditions of impotency among indigenous workers, the mere existence of a foreign power structure can wreak subtle havoc. It does not take a psychologist to perceive that ego strength or views of self-worth and capacity for initiative are not likely to be fostered in foreign enterprises where, as in the case of some U.S. oil companies in Latin America, all employees are required to speak English and the foreign boss, after twenty years in the country, cannot speak ten words of Spanish. I feel certain the situation is no different in the Arab countries of North Africa and the Middle East.

In many respects, then, the imperialists' record in the underdeveloped countries is a bad one. Yet well-meaning but naive people often still argue that the international oil companies made it possible for the oil-exporting underdeveloped countries to develop, and that the failure of these countries to do so was the failure of their own social systems. They concede that foreign companies have taken enormous amounts of capital out of the countries, but urge the point that, in recent years at least, the governments did get sizable sums which they largely dissipated. Second, they note that the social systems of many of the oil-producing countries are among the most backward in the world. In Saudi Arabia, to take the most notable example, slavery was only abolished relatively recently, and a thief can still be punished by having a hand chopped off.

What this line of argument fails to deal with is that, as we have mentioned, the backward and feudal social systems of the oil-exporting countries were strengthened, often deliberately, by the international oil companies and their home governments. Thus, the "windfall" revenues from oil have not generally served as a modernizing force to propel the country from feudalism to capitalism, as happened historically in the developed capitalist countries (whether the windfall at that time was gold finds or plunder from war or slavery). Instead, they have been channeled into preserving the status quo, or dissipated in wasteful consumption, which is itself a way of preserving the status quo.

For the companies, it was much better that government oil revenues be used for importing Cadillacs and air-conditioned palaces and military equipment than for internal development. For one thing, internal development would increase the general demand for labor and hence might force the companies to pay higher salaries. (It is interesting to note that strikes by oil workers in Saudi Arabia are strictly forbidden by law, and punishable by heavy penalties.) More important, economic development might raise the general educational and political level of the people, leading them to ask why the companies should be allowed to extract such enormous profits and have such an important say in the life of the country. The corollary of this is that economic development might undermine the power of the feudal sheiks and corrupt generals—the "strong men"—whom the companies favored because they could usually be bribed with such a small piece of the oil bonanza. And the home governments of the companies were only too willing to assist in this process of preserving the status quo by alternately offering the carrot of foreign aid, particularly military aid, and brandishing the stick of potential intervention.

In retrospect, the highwater mark of Western power and control in the oil-exporting countries probably came in 1953, the year of the Anglo-American coup which denationalized the Iranian oil industry and restored the Shah to his throne. After that point political changes began to take place which would ultimately weaken Western power and hence Western economic control.

One critical set of events clearly was Gamal Abdel Nasser's successful nationalization of the Suez Canal in 1956, as well as the failure of the French-British-Israeli attempt to overthrow him; this not only generated a rift in the Western alliance, but catapulted the Soviets into an important position in the area. Another landmark was the 1958 overthrow of King Faisal in Iraq, which led to the virtual breakup of Dulles's CENTO alliance as a mechanism of Western dominance in the area. Following that, the Cuban Revolution in 1959 shook the American empire, as Fidel Castro made a triumphant tour of Venezuela and successfully beat back an attempt to strangle Cuba with a Western oil boycott. And finally, the Organization of Petroleum Exporting Countries (OPEC), which was to become a household name after the 1973 October War, got off to a shaky start in mid-1960.

The ironic point is that for most of its life, up until the 1970s, OPEC had only mixed success, even in terms of achieving its own limited goals. It was formed in 1960 in direct response to cuts in the posted prices for crude oil which reduced the revenues received by the governments of the oil-exporting countries. The founders were Venezuela and the four leading Middle Eastern producers; in subsequent years all major oil-exporting underdeveloped countries have also joined. The general approach of OPEC in the 1960s has been described by David Hirst as follows:

> Above all, OPEC has always sought to justify its existence through its own competence and expertise. . . .
> [The OPEC technocrats'] appeal to public opinion was from the beginning essentially apolitical, in that they did not identify themselves with any particular political group or interest in their home countries. Their sole object, were it in revolutionary Iraq or absolutist Saudi Arabia, was always to improve their country's "take" from the oil industry. This involved, above all, a thorough

understanding of the workings of this industry and an ability to challenge the companies on specific issues.

In pursuing its general goal of "improving the take" OPEC sought a variety of specific measures. These included prevention of further cuts in posted prices for crude oil, restoration of older higher levels of posted crude oil prices, and increased profit shares for the governments. In addition, OPEC sought greater national participation in the oil industry through increased employment for nationals, the building of new refineries within the producing countries, and government equity in the industry.

In general, it may be said that OPEC's greatest success in the 1960s came in areas where the companies traditionally have been most willing to make concessions to the individual governments; conversely, where OPEC's demands most threatened the vital interests of the companies, it had virtually no success. Specifically, after the formation of OPEC there were no further cuts in posted prices on which producing-government taxes were based (although market prices continued to decline), and while this may perhaps be attributed to the formation of OPEC, it is also true that the companies had not lost their freedom of actual market pricing. The success of OPEC's stand here had effectively been to increase the governments' share of oil revenues, without formal recognition of such a change, while still leaving the individual governments as essentially passive tax collectors. Similarly, OPEC's other main success, negotiating changes in the tax treatment of royalties, had also served to increase the governments' revenues as a tax collector.

On the other hand, in the 1960s OPEC had failed to get posted crude oil prices restored to their former high levels. Moreover, in perhaps the most critical question, national participation in the oil industry, OPEC had relatively little success with the established majors; as one analyst noted in the mid-1960s, "except for the Shell group, all the international major companies have set their face against any deals involving sizable government participation." Again, OPEC failed in the field of obtaining specific price information from the companies:

> ... the major companies agreed to consider a further reduction after 1966 of the discount for tax purposes, taking into account such evidence of the state of the market as the host governments could put before them. The companies, therefore, received letters from the governments asking for details of the prices paid on their actual sales of oil during 1964 and for continuing data. The majors answered negatively since, as they say, they did not promise to provide price details.

Furthermore, despite high hopes of obtaining governmental participation in all levels of the international oil industry, little success had been achieved in this area. Plans for governmental marketing of oil abroad, ownership of tanker fleets, and control of overseas refining and marketing facilities either fell by the wayside or were carried out on a token basis.

OPEC's relative lack of success in achieving major gains had called into question its own existence:

> After OPEC had been in existence for three years, and—apart from the very real but not very inspiring achievement of preventing a further decline in prices—had little concrete to show for it, it was natural that voices should begin to be heard questioning its methods and outlook. ...
>
> ... Ironically, it was OPEC's chief architect, Shaikh 'Abd Allah Al-Tariki—who, in a theatrical outburst, called OPEC the "daughter of the [Arab Petroleum] Congress" and went on to invite his audience to commit infanticide. OPEC, he said, had been wasting too much time on studies and negotiations while the oil companies had been collecting profits to which the Arab people were entitled. ... Tariki, the former advocate of moderate and scientific methods, now outdid Egypt in the popular vein: "Don't depend on the agreements," he warned the companies, "the governments are under pressure; if the people don't like the agreements they'll tear them apart with no compensation."

The basic inability of OPEC to obtain fundamental changes in the position of the oil-exporting governments in

the 1960s may be summarized as due to the following factors. First, the companies, true to their profit-seeking obligations to their stockholders, would not readily concede any
points where their vital interests, particularly control over
prices and production, were at stake. The companies would
only yield where there was a real danger of being forced into
a worse situation if they did not. Second, OPEC's great emphasis on negotiation and world opinion was interpreted by
the companies as a sign of weakness. The companies had
amply demonstrated that they would not hesitate to use
force (oil boycotts) or appeal to their home governments'
force to protect their vital interests vis-à-vis the oil-exporting
countries; OPEC's virtual abjuration of force or nationalization could logically be seen as stemming from a position of
weakness. (When combined with the high time-value of
money to the companies this situation made stalling on their
part a highly profitable business.) Third, OPEC's weakness
seemed to reflect basic divisions and lack of solidarity among
the OPEC member countries. The divisions between the
OPEC members, and particularly the conservative nature of
most of the member regimes, in turn virtually forced OPEC
itself to take a moderate approach.

As a result of all these factors, OPEC increasingly looked
like an organization which was of more long-run value to the
oil companies than to the oil-exporting countries. While this
may appear paradoxical, it is analogous to the theory that
trade unions, properly moderate, play a sufficiently useful
role in providing labor force "stability" to outweigh the
losses to employers from paying out increased benefits. This
view was beginning to prevail among the oil companies, as
indicated by *Petroleum Intelligence Weekly*'s report on the
Sixth Arab Oil Congress held in Baghdad in 1967: "Amid
Baghdad's changing facets, OPEC seemed an element of stability to many oilmen, much as they'd huffed and puffed
about its existence only a few years ago."

OPEC, as it operated in the 1960s, had utilized moderate
methods and obtained only marginal gains. At the same time,
however, it had provided a rallying point and an outlet for

voicing some of the aspirations of the oil-exporting countries. The danger for it, and for the international oil companies, was that OPEC would appear ultimately impotent and lose all influence:

> In appealing to world opinion, OPEC runs essentially the same risk as it does in appealing to a moderate body of opinion in the Arab world itself. That is to say, it inevitably strikes a note of patient exhortation, one might almost say entreaty, which tends to jar on those more nationalist in outlook, who believe that the Arabs and all in their situation should adopt forceful methods as a matter of sovereign right.

This danger was implied by OPEC itself in a 1965 paper analyzing the organization's experience with the oil companies in the drawn-out negotiations over "royalty expensing," an issue which involved a few cents a barrel in the division of profits:

> ... a brief examination of the development of the OPEC royalty negotiations was attempted with particular emphasis on the practices and procedures employed by the oil companies. Generally, the tactics used include: (a) delaying tactics; (b) tactics to undermine OPEC collective bargaining; (c) the making of obviously unacceptable offers; (d) attempts to divide Member Countries; and (e) last-minute modifications of offers before OPEC Conferences.
> In view of OPEC's own experience of negotiating with the major oil companies, the following conclusions may be drawn: ... the major oil companies, by their own behavior, have made the principle of negotiations, which OPEC generally endorses, increasingly hard to apply.... Therefore, this experience automatically opened an unavoidable debate within OPEC regarding the very principle of negotiation itself in future dealings with the major oil companies on issues that might be much more important than the royalty issue.

It would appear that OPEC was transformed into a significant factor in the oil area in the 1970s largely by major changes in the international balance of power and in the strategy of the international oil companies. These changes, which were building up all through the 1960s, burst upon an unsuspecting world in the form of the dramatic developments of the 1970s that made the "energy crisis" a household word. In a later chapter we will discuss these developments and also examine the likely future relationships between the OPEC countries and the rest of the world. At this point, however, we want to analyze the prospects for economic development within this group of nations, particularly in the Middle East, where it appears that huge sums will accrue from oil revenues.

Already the revenues of OPEC governments have increased dramatically, from $6 billion in 1969 to $23 billion in 1973, of which the Middle Eastern countries got $14 billion. Moreover, according to authoritative estimates, OPEC revenues will catapult to close to $100 billion in 1974, of which Saudi Arabia alone would get perhaps $20 billion, Iran $15 billion, and Kuwait $10 billion.

Because of the inflow of such great revenues, under current social, economic, and political conditions prospects for economic growth, in the sense of increased economic activity and GNP, are good. However, in my view, the prospects for real economic development, in the sense of a rapidly rising standard of living for the great majority of people, are poor. Instead, what is likely to happen is that the increased oil revenues will lead to the handful of rich plus the state bureaucracy getting much richer, while the masses of poor will at best enjoy a marginal increase in living standards.

There are a number of reasons for this dire projection, which rests on the assumption that the current social systems of these countries will continue basically intact. First, much of the increase in oil revenues will go to countries with small populations which cannot utilize the money effectively at home, notably Saudi Arabia. Moreover, the countries which will be the biggest recipients of oil wealth are among the most reactionary, notably Saudi Arabia again, but also Iran and Abu Dhabi.

It is obvious that the reactionary regimes of the area cannot generate real economic development because basically they are run for the benefit of the small number of well-to-do, who are opposed to any widespread increases in income which would undercut their power and social position. Undoubtedly, the vast new oil wealth will lead to some improvement in mass living standards, through government welfare programs in education, health, etc., designed to upgrade the quality of the labor force as well as to head off dangerous discontent. In addition, the oil revenues in these countries will be used partly for a handful of major capital-intensive projects, based on petroleum and natural gas, such as petrochemical plants and steel mills, which will, however, employ only a relative handful within the country.

Another part of the revenues will be used for military equipment and supplies, both to fight off internal revolution and in rivalry with other powers in the area. Thus, a major arms race between Saudi Arabia and Iran has already been triggered by the increased oil revenues. What is left will then be used for investment either in the developed countries (stocks, bonds, or direct investment, as for example by Kuwait in South Carolina real estate) or in neighboring countries which lack capital. Saudi Arabia is already shaping up as an important investor in Egypt, which under Sadat is moving rapidly to the right and opening up to foreign investment.

While these various forms of investment may reap substantial profits for the oil-exporting governments, and for the small cliques which buttress them, they will not generate jobs and increase income on a wide scale for most of the people of the country. Rather, as this kind of narrow modernization takes place there will be rapidly increasing wealth for the merchants and traders, real estate speculators, importers of luxury goods, wealthy landowners, and the top bureaucrats and army officers.

Moreover, as the gap between rich and poor grows wider, the increasing tension in society will require ever greater governmental repression of the people to keep the regimes in power, thereby further stunting real economic development. The result will be that these oil-rich countries will become even more extreme cases of capitalist economic development in the Third World, where temporarily high growth rates in the overall economy are accompanied by increasing inequalities of income distribution and increasing tensions and polarization. As an example, we have this 1974 *New York Times* report on discontent in Indonesia:

> Of all the criticisms of the Government, perhaps the charge of corruption is the most explosive. Many analysts, Indonesian and foreign, think the Government could persuade its critics to curb their expectations and show more patience with the trickle-down effect of the development process if the extravagant life-style of the élite was not so visible.
>
> While trash pickers who live in cardboard lean-tos rummage through the garbage bins of the affluent looking for bottles, cans and bits of plastic to resell for a few pennies, the Government's oil czar holds what was reportedly a million-dollar wedding for his daughter on his multi-villa estate, with closed-circuit color television for those guests who could not fit into the hall for the ceremony—guests whose cornucopia of gifts for the bride and groom were said to include a dozen new cars.

The basic problem is that in virtually all of the OPEC countries real power is not in the hands of the people. It is centered either in the old tribal-based rulers, or in the hands of a new class, consisting largely of leaders of the army, the state bureaucracy, and elements of the well-to-do capitalists. Sustained progress is blocked by the unwillingness of the governments to yield power to the people as a whole, which is the vital prerequisite for tapping the mass energy needed for long-run economic development. Without such changes some OPEC countries could also increasingly become stagnant rentier states clipping oil coupons, which is a very dangerous situation, especially when the asset underlying the coupon is a depleting one.

Does all this mean that the oil-exporting countries are doomed to mass poverty and social backwardness in the midst of great riches? In the long run, at least, I believe the answer is no. Already the Middle East has shown that a most backward feudal nation can be relatively quickly moved in the direction of a progressive and egalitarian society by tapping the energies of the masses of the people. The country where this process is taking place—the little-known land of southern Yemen—is such a fascinating case that it is worth looking at in some detail.

Yemen is a small Moslem territory south of Saudi Arabia, with a population of less than ten million. For over a hundred years, until 1967, the southern part of the territory, with only 1.5 million inhabitants, was ruled by the British; according to Joe Stork, the British "followed a policy which one Yemeni characterized in a conversation as 'applied anthropology': arming and subsidizing ruling families to preserve and accentuate tribal division and perpetuate an archaic and decaying social order." They took direct control of the great port of Aden, which was their sole interest in the territory, to use as a military base and a gateway for dominating commerce in the area of Arabia and the Red Sea. There was virtually no industry in the country, with the exception of British Petroleum's refinery, which was built in the 1950s to produce fuel for the many ships using Aden as a refueling port. Agriculture in the countryside stagnated under the rule of local families, and thus served as a useful source of unskilled labor for the menial jobs open to Yemenis in Aden— ship and dock workers, porters, streetsweepers, small shopkeepers, guards, and drivers. At the same time, the situation was even worse in the northern part of Yemen, where "centuries-old rule under the Imam . . . based on tribal and religious divisions, had kept Yemen as one of the most enclosed, isolated, medievally obscurantist countries in the world."

Three key external events brought the first rumblings of change to southern Yemen—events whose vast impact dramatizes the interrelatedness of the modern world. The first was the overthrow of the old order in Egypt by army officers under Nasser in 1952. Egypt became a place where young Arabs, including Yemenis, could get higher education with government help, and also come into contact with nationalist and radical ideas and literature. At the same time, the change in Egypt forced Britain to pull out from its base in Suez;

this, along with similar withdrawals from Cyprus and Kenya, led to the strengthening of the Aden base, which in turn led to renewed opposition to it within Aden.

The second key event was Mossadegh's nationalization of British Petroleum's Abadan refinery in Iran in 1951, which led the company to build the Aden refinery as a substitute. This in turn led to the inevitable growth of the first fairly skilled and permanent industrial labor force, which was to become a bastion of new revolutionary forces. The third major development was the coup against the Imam in northern Yemen in 1962 and the establishment of the Republic of Yemen there, backed by Egypt. This event not only provided a safe rear base for the liberation forces in southern Yemen, but also helped shake the people out of centuries of apathy by showing that armed action could be successful in the area. If the Imam could be toppled, why not the British?

Soon after the coup, in mid-1963, a National Liberation Front was formed in southern Yemen and in October of that year armed struggle began from the mountains. After four years of such struggle the NLF took power from the British, who were forced to withdraw. Britain, however, was not the only source of difficulty in these four years. Initially Egypt backed the NLF, whose dominant political line was the vague ideology of Nasserism—anti-imperialism and "Arab socialism"—but it also tried to dominate. When frictions developed between the NLF and Egypt, the latter set up an umbrella group in 1966 called the Front for the Liberation of Southern Yemen (FLOSY), from which the NLF broke away. In the brief struggle between the NLF and FLOSY, the NLF won a clear victory. The basic reason was that it had organized the countryside, while FLOSY's only strength was in Aden.

When the NLF took power in 1967, it appeared to be a typical "revolutionary nationalist" regime, such as characterized Egypt, Algeria, and Iraq. In fact, so confident were the British about its long-run moderation that this confidence, combined with their phobias about Nasser, led them to influence the local southern Arabian army to support the NLF over FLOSY. However, it soon became clear that the leaders of the NLF, who were almost all young men in their thirties, were at least open to more radical, Marxist ideas. Thus, the country was named the People's Democratic Republic of Yemen (PDRY). More important, at the National Congress of the Front in early 1968, the Minister of Information "cited the Arab Socialist Union in Egypt, the Ba'ath of Syria and Iraq and the Algerian FLN as examples of 'self-styled socialist parties which are petty-bourgeois in ideology and class make-up.'" Furthermore, at this Congress "resolutions spoke of the need for 'the victory of our revolution in the north' and the need to take up its 'historical responsibility for the Arabian Gulf and all areas of Arabian Peninsula for the elimination of imperialist and reactionary forces.' Internally the Congress called for elimination of 'exploitative relations' in the countryside and ending the domination of foreign capital in Aden."

The difference between the government of the PDRY and other regimes in the area, even the "progressive nationalist" ones, is fundamental. Virtually all major foreign enterprises, except the refinery, have been nationalized—banks, insurance, foreign trade, petroleum marketing, and port services. (The significance of the refinery had already been reduced by the 1967 closing of the Suez Canal, which sharply undercut Aden's position as a refueling port.) Where firms have been left in private hands, it has been for pragmatic reasons, such as the vast economic and administrative problems which would at this stage be involved in government ownership of certain sectors. In addition, the Three-Year Plan (1972-74) involved construction of thirty new public factories, with the aim of diversifying industry away from the dominance of the refinery.

Many of these measures can be found, however, in other "progressive nationalist" countries. What is truly unique here, and much more significant, is the fundamental transformation of the countryside, where 80 percent of the people live. Through a series of "uprisings" the peasants and fishermen of the country have wrested control of the basic means of production—the land and the fishing ships—from the small group of wealthy families which had previously controlled them and through them the cultural and social life of the country. Backed by the power and authority of the state, and stimulated by young workers and students, the peasants and fishermen formed militia to challenge and beat the tyranny of the local well-to-do. The land and tools and boats were then taken over by newly formed cooperatives.

As a consequence of this true revolution in economic power, it was possible to make a start on improving social conditions. In the area of health, more clinics were built in three years than in 129 years of British rule. Since 1967 the number of primary schools has jumped fourfold and the number of intermediate schools has more than doubled, with similar increases in numbers of students and teachers. Moreover, the educational system places new emphasis on manual and technical work, in order to build the human basis for economic development.

In the touchstone area of the role of women, considerable progress has been made. Girls now account for almost 20 percent of all primary school students, up from almost zero before the revolution, and it is planned that 75 percent of all girls will be in primary school by 1980. Women are now encouraged to give up the "veil" and work outside the home. (This often meets fierce resistance; in one place, for example, when the national bank employed a woman as secretary—the only woman employed in the area—she became the target of organized slander by the local traditional elite.) As Stork noted in early 1973, these developments

represent the important first step in a society that has been virtually frozen for centuries in this regard. At this point, women are not organized into a strong and viable political organization that could survive and struggle in a politically hostile environment or in the face of regressive measures by the authorities such as happened in Algeria. The important fact is that the National Front has articulated and demonstrated a genuine concern to change the condition of women in the course of building the new society.

The danger to imperialism that southern Yemen might be the "face of the future" for the Middle East has long been recognized. Saudi Arabia has supplied large amounts of troops, money, and materiel to overthrow what its King Faisal has called a "satanic citadel of subversion," and has sought to have the Republic of Yemen in the north conquer the PDRY in the south. Again, Britain and the United States have been seeking to contain the spread of the southern Yemen example as embodied in the Popular Front for the Liberation of Oman and the Arabian Gulf (PFLOAG), which is carrying on revolutionary war in that neighboring area. The efforts of the PFLOAG are supported politically by the PDRY, which also provides transit facilities for outside supplies. However, the PFLOAG is opposed far more blatantly and massively by the troops and planes of the Western junior partners in the area, particularly Pakistan, Jordan, and most important, Iran, which has invaded the area with 3,000 troops. Thus, the forces of revolution in the Middle East face a difficult struggle over the coming years.

However, what seems certain is that in the long run there will be increasing tension between the established forces of the status quo and of "progressive nationalism," on the one hand, and radical revolutionaries on the other. The basic dilemma for the existing regimes is that at best they will generate economic growth and not development, and that as the gap between what could be and what is done with oil revenues grows, popular dissatisfaction will increase.

At the same time, tensions will be heightened by frictions and struggles between the developed countries and the local regimes. All this will be further exacerbated by increased rivalries among the major developed countries for oil resources. To see why this is likely, we turn in the next chapter to an examination of the role of those major developed countries which are not at present major oil powers on the order of the United States and Great Britain.

5

Western Europe and Japan:
The Losers

Reports of mammoth deals hastily being negotiated between Western European powers and the oil-exporting countries to exchange sophisticated military equipment for crude oil emphasize once again how success in the twentieth century international "oil war" depends upon military as well as economic power. This will be underlined here as we examine the historical failure of Western Europe and Japan to break the Anglo-American stranglehold on the industry, as well as the steps they are now taking to deal with their exacerbated problems.

The classic example of "finders keepers, losers weepers" in the oil industry is Germany. In the last part of the nineteenth century rising German power was already reaching out in the oil area, seeking fuel for its military and industrial machine, with the Deutsche Bank of Berlin as its principal instrument. The Bank bought one-half ownership in Steau Romana, a major Rumanian oil producer whose output had been going to Royal Dutch, as a step toward breaking the Standard Oil monopoly within the German market. After a fierce struggle, including a price war, agreement was reached in which Standard was limited to "only" 80 percent of the German market. The Deutsche Bank tried other methods to break Standard's near monopoly of its home market. Before World War I it tried to buy up Texaco's and Gulf's American surplus oil, and even Gulf itself—for $25 million—but was blocked by Standard's purchase of these companies' export surplus.

As a relative latecomer to the empire business, Kaiser Wilhelm's Germany had been boxed out of most of the attractive parts of Asia and Africa (not to say Latin America, which of course was Uncle Sam's turf). Therefore, it eyed the Middle East with particular lust.

Around the turn of the century German experts had described Mesopotamia as a "lake of petroleum," "virtually soaked with bitumen, naphtha, and gaseous hydrocarbons." As one step toward gaining access to this, Germany had set up a railroad from the Bosphorus to Ankara intended as part of a Berlin-Baghdad-Bombay backbone for its empire. It then began dickering to obtain a terminus for the railroad in the obscure Persian Gulf port of Kuwait, but was rebuffed by British power; as one writer put it, "Lord Curzon, then Under-Secretary for India, announced with some asperity that the British lion would take care of the sheikh and his sheikhdom."

Nevertheless, the Deutsche Bank continued to maneuver for the rich Mesopotamian prize, and in 1904 was given the right to make a survey of its resources. The jockeying for position in Mesopotamia, part of the Turkish Empire, was involved and complex, but finally led to the setting up of the Turkish Petroleum Company in 1914, with the Deutsche Bank having a one-fourth share, one-fourth going to Royal Dutch Shell and one-half to British Petroleum. The Turkish Petroleum Company's claims to a concession for the area were disputed by the Turkish government, but the outbreak of World War I changed everything anyway.

As we have seen earlier, the rivalry continued for many years after the war, and was not finally settled until 1928. However, for Germany the issue was settled decisively by World War I, because as a loser its one-fourth share in the Turkish Petroleum Company was turned over to a victor, France, at the San Remo conference of 1920.

As a beaten and bankrupt country, Germany after World War I was not in a position to do anything about gaining a foothold in the oil area. Hitler, of course, tried to change this situation during World War II by driving to take over Eastern Europe and the Middle East, but his ultimate failure left Germany even weaker. Since the United States became the dominant power after the war, West Germany as an occupied country was hardly in a position to tangle with Anglo-American oil power abroad, or even to resist the forced switchover from its bounteous indigenous coal reserves to imported oil. Even today West Germany has the weakest oil position of any developed country, and only in 1970 did its government finally assist in setting up a German-owned company to seek oil abroad.

France, as an ultimate victor in the two world wars, has clearly fared better than Germany. The latter's quarter-share in Iraq's oil, which was given to France after World War I, formed the basis for the CFP, in which the French government retained a controlling interest. Through its interest in the Iraq Petroleum Company, CFP also obtained quarter-shares of concessions in Qatar and Abu Dhabi in the Persian Gulf. And, of course, when the Iranian melon was being divided up after the Anglo-American counter-revolution in 1953, because CFP was a significant Middle Eastern factor it was able to get a 6 percent interest there.

CFP was further strengthened by the French government, which from the earliest days of the petroleum industry had taken a very interventionist role, by being awarded in 1929 one-fourth of the French market for refined products. But the most important role played by the French government was that, as a victor in the two world wars, it retained colonies, and later neocolonial influence, which provided both crude oil supplies and markets for French-controlled refined products.

By far the most important of these colonies was Algeria, where oil was found in the mid-1950s and was a major factor in France's stubborn unwillingness to agree to independence for the country. It was to explore for oil in Algeria that Entreprise des Recherches et d'Activités Pétrolières (ERAP), a wholly French government-owned company, was formed. ERAP was the major winner in the Algerian oil sweepstakes, while CFP was in second place. ERAP and CFP also have small amounts of crude oil production and refining capacity in many of the former colonies of French West Africa (Congo, Gabon, Ivory Coast, Senegal, etc.).

Despite the success of France in getting an oil foothold in a number of countries, its Achilles heel has been that it was just that—a foothold. French companies have never had a major position in any of the real bonanzas of the oil world, and in the one place which looked promising—Algeria—they lost much of their position to nationalization. As a result, French crude oil production has not been nearly large enough to satisfy its rapidly growing oil consumption after World War II. Today, crude oil supplies controlled by French companies account for less than two-thirds of French crude oil consumption.

Partly because of this costly deficit situation, in recent years France has tended to come into conflict with the Anglo-American monopoly. Its historic link with these companies through CFP's position as the junior member of the Big Eight has been greatly weakened as France has sought to enlarge its foothold in the oil-producing world. The most dramatic example of this centers in the international power struggle in the 1960s as to who was to develop the rich Rumalia field in Iraq.

The stage for this was set back in 1961 when the Iraqi government took back from IPC—the Iraq Petroleum Company, a consortium of international companies which controlled Iraq's oil—the concession for production in areas not then being exploited. This revoked concession area covered over 99 percent of the country, including the North Rumalia field which was estimated to have reserves of five billion barrels. A dispute between the companies and the Iraqi government over this action simmered for years while oil exploration in Iraq was at a standstill. Then, in early 1967, ENI (Italy's state oil company) entered into negotiations with the Iraqi government to exploit the disputed area, which in turn led the governments of Great Britain, the Netherlands, and France to make diplomatic "representations" on behalf of their oil companies (CFP being a member of IPC) to the Italian government. Not to be outdone, the U.S. State Department, according to government records:

> leaned on 10 U.S. companies, asking each not to sign agreements with the Iraqi government, pending settlement of the concession dispute. Moreover, top executives of Jersey Standard [Exxon] and Mobil asked Assistant Secretary Anthony Solomon to "threaten economic retaliation," even against friendly foreign governments, to stop competitors from muscling in on the consortium area. ENI, the Italian government corporation, was thought to be the chief intruder.
>
> The State Department complied. "After careful consideration of the dangers of alienating the government of Italy, we have gone to great lengths to support Iraq Petroleum Co. against depredations by ENI," a department document says.

In late 1967 France mounted a two-pronged attack to enlarge its position in Iraq. First, the wholly state-owned ERAP signed an agreement to explore all unproven areas for Iraq's state oil company. Soon after, CFP itself began separate negotiations with the Iraqi government with the aim of exploiting the North Rumalia plum. This latter step particularly set the international oil pot to boiling, as indicated by a report in *Petroleum Intelligence Weekly:*

> Most of CFP's partners in Iraq—British Petroleum, Shell, Esso and Mobil—are indignant, to say the least. They consider the CFP move as an "unauthorized breach" of the Iraq Petroleum Co. convention and both the U.S. and British Governments have protested strongly to the French Government.
>
> In Paris, diplomatic sources confirm CFP's claim that its action has been taken at the urging of the French Government (which owns 35% of CFP and 40% of its voting rights). The government maintains IPC has no chance of getting back North Rumalia, and it doesn't want France to lose out to the Italians, Japanese and other potential bidders. And the move, moreover, fits with France's overall policy of trying to augment its interests in Middle East oil.

A columnist for the *Oil and Gas Journal* commented with skepticism about CFP's justification of its move in terms of preemptive necessity, noting that "this was the same alibi used last spring by Italy's ENI when it first came to light that ENI was trying to make a similar deal with Iraq." More important, the writer stated, was the indication "that de Gaulle has now decided that France can go it alone in oil affairs, and no longer needs to be cautious about stepping on Anglo-American toes."

France's activities in Iraq may in retrospect be seen as a historic milestone in international oil history. The possible results were clearly forecast by John Buckley, executive editor of *Petroleum Intelligence Weekly*, in a 1967 talk to the New York Society of Security Analysts, which was summarized in the *New York Times* as follows:

> ... [Mr. Buckley] believes France may have shattered the foundations of the long-standing arrangements under which the international oil companies pump black gold throughout the world.. ...
>
> He said that when France's state-owned Entreprise de Recherches et d'Activités Pétrolières took over confiscated concessions in Iraq, it broke faith with other consuming nations which had previously refused to support similar seizures by producing Governments.
>
> E.R.A.P. has agreed to finance a search for oil in concession lands once owned by Iraq Petroleum Company. Mr. Buckley said that if France does this, other Arab nations are likely to break concession agreements that permit oil companies to reap big rewards. ...
>
> Mr. Buckley said that the United States is now in danger of losing its control of Arab oil, which has put the nation in a very strong international position.

Ultimately, after a 1968 coup, the new Iraqi government decided to develop the Rumalia field for itself, with Soviet assistance.

How right Mr. Buckley was about France's role was revealed when Iraq nationalized part of the holdings of IPC in 1972, and an attempt was made to organize a Western boycott of oil from these fields, similar to the boycott which had helped overthrow the Mossadegh government in Iran twenty years earlier. While the Soviet Union played the most crucial role in undercutting this boycott, both by agreeing to buy large amounts of Iraqi crude and by providing major military and economic assistance, France also helped by undertaking to buy large quantities of Iraqi crude on a long-term basis. Thus, despite the fact that CFP owned almost one-fourth of the nationalized oil fields, France's overall interest lay in conciliation rather than confrontation. And, it might be noted, the rewards for this were not long in coming, since the Iraqi government's nationalization of all U.S. and Dutch interests in IPC soon after the 1973 October War left the French in a favorable position there. Today, Iraq has surpassed Algeria as the leading supplier of French-controlled crude oil.

Italy is the one other major Western European country which has played a significant role in the international oil industry. This is principally because of the post-World War II activities of ENI, under the leadership of the late Enrico Mattei.

Historically, Italy had been handicapped by a lack of indigenous energy resources of any kind, except for some hydroelectric power. Moreover, as a weak colonizing power, even though it was a victor in World War I it could not get any of Germany's share of the potential oil reserves of Iraq, which went to France instead.

As a first step toward rectifying this situation, Mussolini's fascist regime in 1926 formed a state-owned company, AGIP (a forerunner of ENI), to explore for oil at home and abroad. Before World War II it had little success, getting only some medium-sized companies in Rumania and access to some Albanian crude oil. A toehold in Middle Eastern oil was obtained through ownership of a company with a concession in Iraq, but even this was lost. Mussolini's clash with Britain over his invasion of Ethiopia made the British eager to ease Italy out of this critical area, and since Italy lacked the financial wherewithal to make a big exploration effort, AGIP was ordered to accept IPC's offer to buy it out.

AGIP did, however, build up a strong distribution network for petroleum products within Italy, and by 1939 had one-quarter of the market (its main rivals were Exxon and Royal Dutch Shell). In addition, another branch of ENI built up some refining capacity in the country. What was lacking, above all, was crude oil. While Mattei, who took over the leadership of ENI in 1946, found sizable amounts of natural gas in the Po Valley, he did not find equivalent crude oil. It was his vigorous drive to obtain foreign crude oil, and the repercussions it caused, that make Mattei's ENI worth examining.

Mattei expressed his philosophy as follows:

> This fresh outlook of ENI's accepts the realities of the new political situation created by the emergence of the underdeveloped countries from their state of political subjection into one of independence.
>
> From this aspect I consider ENI's formula not so much as a means for a company to penetrate into some of the most sought after oil-bearing regions, but rather as an initial step forward to more lasting relationships between oil producing and oil consuming countries. . . .
>
> As the ultimate responsibility for oil operations passes gradually over to the state, whether producer or consumer, it is the states themselves which are becoming the real protagonists and, as the chances of a compromise of interests within the framework of the agreement between the major international companies fade, favorable circumstances are being created for setting up a new system based on co-operation between producer and consumer nations.

Under Mattei, ENI, whose intended function was to provide low-cost energy for Italy, played the dual role of attempting to beat down the price of imported crude oil, and at the same time gain for Italy direct ownership of overseas crude oil by providing more attractive profit-sharing deals with the governments of the oil-producing countries. Thus in the 1950s ENI became a purchaser of large amounts of low-priced "barter oil" from the Soviet Union. As a seeker after overseas oil, it shattered the standard 50–50 profit split among the oil companies and the governments of the oil-producing countries by offering a 75–25 split in favor of the governments. At the same time, both as a crusader against the power of the international oil companies and for its own profits, ENI offered better deals to the oil-importing underdeveloped countries:

> ... whereas other oil interests built refineries to secure outlets for their crude and/or to protect their distribution networks, ENI, not having crude oil of its own, managed to get favorable consideration for its proposals mainly because it could consider refining as a business on its own ... apart from an attractive return on capital, ENI usually had the benefit of design and engineering contracts placed with its affiliates and, having got a foot in the door, AGIP [an ENI affiliate] also managed to establish itself as an oil distributor in the country.
>
> All these activities in the underdeveloped countries, although some of them were more spectacular than substantial, were on the whole and with few exceptions sound and not unreasonable in terms of profitability.

Unfortunately for Italy, Mattei never found much crude oil abroad (he was called by his critics "the oilman without oil"), even while the country's oil consumption expanded enormously. Thus, by 1962, when Mattei died in a suspicious plane crash, ENI produced only a minor part of Italy's oil consumption. After Mattei's death ENI seems to have become much more conciliatory in its dealings with the big international oil companies. As *Petroleum Intelligence Weekly* noted in 1965:

> In the final analysis, economics don't tell the whole story behind the decision of Italy's state-owned ENI to sell out its properties in Britain to the world's biggest private oil company, Esso [Exxon]. . . .
>
> ENI's U.K. company, like the one in the U.S., was set up in 1961 by Enrico Mattei, at least partly as a response to an emotional drive to establish "a beach-head" in the home territories of the major international companies—called derisively by Mattei, "the seven sisters". . . .
>
> Since Mattei's death, his successor, Eugenio Cefis, has established a working rapport with the big internationals and crude oil purchase deals have been negotiated by the crude-short Italian company with both Esso and Gulf Oil. At the same time, the U.S.

affiliate, "Agip, U.S.A." has been confined primarily to supervising material and equipment purchases, despite earlier plans to move aggressively into Western Hemisphere markets. And the U.K. Company has expanded much less quickly than ENI had initially expected.

Because ENI has failed also in the last ten years to find large amounts of crude oil, it now produces less than one-sixth of Italy's enormously increased consumption. Hence the country, which is also plagued by a chronically weak currency, is particularly caught in a squeeze. As we shall discuss later, because of the vastly increased oil prices, Italy is trapped between the forces of the Big Seven and their home governments, on the one hand, and the oil-producing countries on the other. This makes it virtually impossible for Italy to be more than a passive spectator as the events unfold which will affect the very lifeblood of its economy.

Japan is the other important developed country whose position in the international oil industry very much resembles that of Italy and Germany. Historically Japan, too, has lacked significant amounts of indigenous crude oil. Its energy economy ran primarily on coal and hydroelectric power, plus oil imported from the major international oil companies. Before World War II Japan too was blocked out from any of the major oil-producing areas of the world. Indeed, like the other Axis powers, Japan's desire to obtain its own supply of crude oil was one of its principal motives in entering World War II. Japan's seizure during the war of Dutch East Indies and Burma oil fields did not help, however, because of British and Dutch destruction of the fields as well as the heavy toll taken by American submarines of tankers transporting the remaining oil to Japan.

After the war, with the American military occupation, the international oil companies took over with a vengeance. They turned Japan, like Western Europe, from a primarily coal-consuming country to an oil-based one, with oil's share of total energy rising from less than one-tenth in 1950 to over two-thirds today. The companies collaborated with Japanese interests in building up a huge indigenous refinery industry since they were primarily interested in Japan as an outlet for their low-cost and high-profit crude oil; the dominance of American companies is shown by their supplying, in 1966, 63 percent of Japan's crude oil, compared to a 15 percent share for English companies.

With the country's ever increasing dependence on imported crude oil, Japan made efforts to gain its own concessions in the important oil-producing areas of the world. The most impressive achievement was the concession gained in 1958 by a privately owned Japanese company (Arabian Oil) from the governments of Saudi Arabia and Kuwait to explore in their "Neutral Zone." This was obtained by offering better terms to the governments than the traditional 50–50 profit split. The Japanese added insult to injury of the Big Seven when they soon found oil in the concession and began, with the assistance of the Japanese government, shipping it to Japanese refineries in competition with Big Seven oil. However, the relatively poor quality of this oil (it has a high sulfur content), combined with the rapidly swelling Japanese demand for oil, insured the continued dominance of the Big Seven in supplying Japan's crude oil needs.

In a real sense, then, the developed countries discussed in this chapter are the losers in the international oil struggle which has been going on for almost a century. These are the countries which, whether owing to lost wars or relative economic and military weakness, have been forced to rely on foreign-controlled oil supplies.

In the 1960s, however, as oil rapidly became the preeminent fuel for their economies, the real significance of this position of weakness was masked by low and falling prices for crude oil. Thus, as the economies of these countries expanded in the long boom of the 1960s, and their exports increased even more rapidly than domestic production, the rising oil import bill could easily be financed. Oil imports amounted to a manageable 10–15 percent of total export revenues for these developed countries.

During this same period these "loser" countries in oil were becoming winners in another titanic struggle with the Anglo-

American world, this time in the international monetary arena. Because events in the international oil arena now are so closely linked with those in the international monetary arena, it is necessary to briefly explain the underlying significance of this monetary struggle.

During the 1960s and early 1970s the capitalist world was rocked by a series of monetary crises, involving fierce struggles over the relative values of dollars, pounds, marks, yen, francs, and lire. In my view, at the core of these recurring monetary crises was the struggle over the role of gold versus the U.S. dollar in the world economy. Underlying this struggle in turn was a fierce economic and political power struggle between Western Europe (and Japan) and the United States over the role of American investment in these countries.

In order to understand the nature and significance of the present international monetary crisis, some historical perspective is needed. In the 1930s, as a result of the Great Depression and the specter of war, capital fled from Europe to the United States, which built up an enormous gold hoard. This stock of gold rose from $10 billion in 1935 to $22 billion in 1940, at which time the United States held the bulk of the world's gold. World War II physically destroyed the economies of all the European powers but Great Britain, which it vastly weakened. At the same time, the economic and financial power of the United States was greatly increased; its gold holdings reached a peak of $25 billion in 1949.

The Marshall Plan helped to rebuild Europe, but partly, as we have seen, at the price of increasing U.S. penetration of Europe and former European colonies. The book value (basically the original cost) of direct U.S. foreign private investment rose worldwide from $12 billion in 1950—the order of magnitude at which U.S. overseas investment stood in the 1930s—to $33 billion in 1960, and to over $100 billion today. And everyone agrees that these figures greatly understate the real value of the assets the United States has accumulated. (As an example, the whole of U.S. Middle Eastern oil investment is carried on the books at under $2 billion, but this figure excludes the value of the hundreds of billions of barrels of oil proven to be in the ground.) Thus, what these figures really show is not the absolute level of investment, but its rapid rate of growth.

At the same time, in almost every year since 1950, the United States had a balance of payments deficit. This means that, including our foreign investment overseas, we have spent more abroad than foreigners have spent here. How have we paid for this deficit? The answer, of course, is to be found in the historic willingness of foreigners to accept dollars or gold in exchange for giving the United States real economic assets. Until 1957 foreigners were willing to take dollars, and the U.S. gold hoard stayed at about the 1950 level. From 1958 on, however, foreigners began increasingly to demand gold, and the U.S. gold stock shrunk to about $11 billion in 1971.

Why the shift? One reason is that, despite U.S. assertions to the contrary, gold is still the only universally accepted currency. The ultimate strength of the dollar as an international currency lay in the huge supply of gold available to reclaim any unwanted dollars presented at the U.S. bank window. As the supply of dollars in the hands of foreigners grew increasingly larger than the value of U.S. holdings in gold, the creditors naturally became increasingly suspicious of our ability to pay their claims in the only currency they really trusted.

A second reason was the drive, particularly by the French under De Gaulle, to slow down U.S. investment in Europe. This investment had risen from $2 billion in 1950 to $6 billion in 1960 and to about $25 billion in 1970. In these two decades the increasingly dominant factor in U.S.-European economic relations had been the growth of U.S. foreign investment. This growth was, in my view, the basic cause of the dollar outflow which, as these dollars were cashed for gold, led to the gold and currency crisis.

However, by the later 1960s the non–Anglo-American developed countries were beginning to enjoy some returns from the power they derived from their vast gold and dollar hoards. It has been argued, for example, that Europe's threat of a run on America's gold hoard in 1968 was an important factor in forcing Lyndon Johnson to call a halt to the bombing of North Vietnam. More tangibly, the weakness of the U.S. dollar which in 1971 forced Nixon to devalue it (and to refuse to sell gold for dollars) made other currencies more valuable. This gave other countries a greater quantity of goods for the same amount of their money, and also enabled them to begin to reverse their relationship with America by investing heavily in the United States itself.

All of this was reversed almost overnight after the 1973 October War when oil prices skyrocketed. Suddenly the oil-deficit position of the non–Anglo-American developed countries was revealed as their Achilles heel. Because the United States imported relatively less oil than these countries, and also had the greatly increased profits of its international oil companies, international speculators and investors rushed to dump their marks, yen, and francs in favor of dollars. Only large-scale borrowing by Western Europe and Japan prevented their foreign exchange reserves, which had been painfully built up over more than two decades, from gushing like blood from a punctured artery.

Moreover, the sums of money which these developed countries would be required to pay for their oil imports at the new price levels threatens to bankrupt them in not too long a period. This has put these countries in a precarious situation from which they are frantically seeking an escape.

What routes they will take in this quest will have a major impact on the international economy for many years to come. However, before we can discuss the choices that these developed countries face, we first have to examine some of the other forces in the international scene which act as constraints on these options. Hence, in the next chapter we turn to an examination of the Communist countries' role in the international oil arena.

6

The Communist Countries

The historical experience of the major Communist countries in developing their energy resources is of considerable interest for at least two reasons. First, it shows how an underdeveloped country's energy resource base can be transformed by the move from a "free market" system to a "planned economy," and the implacable enmity which such a move generates on the part of the international oil companies and their home governments. Second, it shows how, since energy resources are so basic a component of a modern society, decisions about their development are inextricably bound up with overall decisions about the kind of society to be built internally as well as the nation's role on the international scene. Therefore, the examination here of energy in the Soviet Union and China should not only shed light on the likely future role of these countries vis-à-vis the energy crisis, but should also be a useful prelude to our analyses of the position of the oil-importing underdeveloped countries and the possible role of energy in a more rational world.

In Czarist Russia the first sizable oil industry had been developed at Baku in the 1870s by the Nobel family (of dynamite and prize fame). The Rothschilds then got a foothold by obtaining a concession to handle Russia's export surplus, and soon became second to the Nobels in production and refining. As we have seen earlier, the Nobels and Rothschilds, on the basis of rapidly rising Russian production, ended Standard Oil's monopoly in Western Europe by the late 1880s, and by the end of the century Russia was the world's largest oil producer, with the United States second; each exported about half of its production. In the next fifteen years, however, Russian production stagnated, while that of the United States, catapulted by Spindletop, more than quadrupled, so that by 1915 Russia's share of world production had dropped to one-sixth.

One of the first acts of the Bolshevik Revolution was to nationalize the country's oil industry. But Russian oil had been viewed as the great prize of World War I, particularly since it was believed that Russian reserves were greater than those of the United States. According to Leonard Fanning of the American Petroleum Institute, by 1930 the United States would be needing Russian as well as Persian oil. Hence, the Nobels and Rothschilds and Standard and the other international oil companies sought to regain their interests in any way possible. The White Russians who fought the Bolsheviks for several years were strongly supported by British, U.S., and Japanese armies, with the former actually occupying the Caucasus until 1919.

When the efforts at violent "recovery" of Russian oil failed, Standard, Royal Dutch and the Nobels agreed in 1922 to boycott Soviet oil—the first major oil boycott in history. The boycott ultimately failed because of shortages of oil as well as squabbling among the participants in the "united front." Nevertheless, it laid the groundwork for the international oil companies' continuing vehement opposition to the Soviet Union, as the carrier of the virulent disease of nationalization.

World War I and the ensuing civil war reduced oil production in the Soviet Union sharply, so that by 1925 it had reached only three-fourths of the prewar level. However, in the five years after that it increased by 150 percent and almost doubled in the next ten. World War II again provided a temporary setback, but by 1949 recovery was complete. By 1961 oil production had more than quadrupled, making the Soviet Union second only to the United States; today the two countries' oil production is very nearly equal.

Despite the impressive increases in Soviet crude oil production, from about ten million tons per year in the pre-World War I period to about 450 million tons per year today, it has always been Soviet policy to build up a balanced energy position. This was necessary partly so that the country would not become heavily dependent upon foreign sources for its vital energy supplies. In addition, stress was placed on utilization of local fuels in order to reduce the strain on the internal transport system. Despite this policy, because the Soviet Union is a vast country and fuel supplies are poorly located in relation to consuming centers, by the early 1960s 40 percent of the country's transport system was required to haul mineral fuels.

In 1928, at the beginning of the First Five-Year Plan, oil accounted for about one-sixth of all energy, coal about one-fourth, and firewood most of the rest. As a result of a Soviet fuel policy aimed at shifting the economy away from the more primitive firewood, by 1950 coal's share of a quadrupled energy production had risen to over half, firewood had dropped to less than one-fourth, and oil to less than one-sixth.

Owing to an enormous increase in oil and gas exploration after World War II, by today when total energy production has increased sixfold over the 1950 level, coal's share has dropped to two-fifths, oil is up to one-third and natural gas one-fifth, while firewood and all other sources have dropped to less than one-tenth.

The Soviet success in the oil and gas area is also shown by the estimates of net trade in energy supplies. In 1928 the Soviet Union exported about 3 percent of its total energy production, but by 1950 this surplus had turned into a deficit of about 3 percent. However, by the early 1970s a surplus of over 10 percent had been reestablished for export.

All in all, as far as production is concerned, the conclusion of one leading Western expert seems fully warranted: "In my opinion, the crucial lesson of the Soviet record for more and less developed countries alike is the productiveness of planning and operations based on physical resources and requirements."

The Soviet Union's success in developing indigenous energy resources, however, is closely related to its successes in the industrialization of the economy as a whole. After all, Lenin had early stated that electrification was the vital prerequisite for developing the economy. Thus, in 1955 (a year for which relatively complete data are available), half of all energy production went to industry, particularly to heavy industry like mineral production and refining, metal products, cement, and chemicals. About one-sixth went to civil transportation, of which three-fourths was for the railways, while another one-sixth was for households and the remaining one-sixth for miscellaneous sectors including agriculture, defense, and construction.

The emphasis of the Soviet planned economy on more efficient public transportation as opposed to private transportation, particularly in the form of the private automobile, has been an important factor in allowing the country to utilize its own energy resources and not rely on imported oil. The quantitative significance of this one factor alone can be seen from comparative data for 1955 for the United States, where total energy consumption was about three times higher, and where 10 percent of this total was needed for private automobiles (including about half of the country's total petroleum consumption). If in 1955 the Soviet Union had had the same level of automobilization as the United States, it would have needed to produce or import about 30 percent more total energy, and more than twice as much in the way of petroleum products.

In this connection, the decisions made by the Soviet leadership in the 1960s to emulate the United States and seek massive private automobilization of the country will undoubtedly have a major impact on energy resource planning. It is not just that the rapidly growing fleet of private autos will require ever increasing amounts of gasoline for a relatively inefficient means of transportation. In addition, vast investments will be required to build up the infrastructure that private cars require, particularly highways, garages, service stations, motels, parking lots, etc., and these too will necessitate large amounts of energy, both for construction and operation.

Moreover, the general decision of Soviet leaders to emphasize production of private consumer goods rather than collective consumer facilities will undoubtedly add to the energy
burden. The decision to increase home washing-machine production, for example, rather than building collective large-
scale laundry facilities, reflects an acceptance of the economic inefficiency of small-scale home equipment in
exchange for the ability to use these items as material incen-
tives for workers and rewards to the new managerial elite.

Thus, even by the early 1970s, when Soviet passenger car
production had risen to over one million per year (or only
one-tenth the U.S. rate), there were some signs of increasing
strain on Soviet oil resources. The ratio of Soviet crude oil
reserves to annual production reportedly dropped from about
20 in 1960 to 14 in 1970, and according to *Petroleum Press
Service* "this ratio apparently continues to deteriorate."
Moreover, Soviet authorities believe that the traditional production areas in European Russia have about reached their
peak, so that increased emphasis has been placed on exploiting the reserves of Siberia.

Because such exploitation would be technically difficult
and highly capital-intensive, the Soviet Union has increasingly
been looking to the West and Japan for financial and technical assistance in this area. If such assistance is not forth-
coming on a large scale, and if the Soviets cannot go it alone,
then access to the vast reserves of the Middle East will clearly
become of vital importance to them. However, since the Middle East is an area of rivalry between the Soviet Union, the
United States, Western Europe, and Japan, it is equally clear
that Soviet decisions in the energy area must be closely related to overall foreign policy considerations.

It is worth noting that in the post-World War II period
there have been two distinct phases as regards the Soviet
Union's foreign energy policy. The first period, 1955–1961,
was the era of an "aggressive" Soviet foreign energy policy.
During this period, Soviet oil exports shot up dramatically,
from eight million to forty million tons per year. This was
accomplished partly by price cutting and partly by offering

barter deals, which were particularly attractive to countries which had foreign exchange shortages.

In this period the Soviets found markets for oil not only in Eastern Europe but in Italy, West Germany, Sweden, Japan, France, Austria, Greece, Egypt, Cuba, and Brazil. At the same time, the Soviets were aggressively offering assistance to governments of underdeveloped countries, in the form of low-interest loans, grants, and technical help, to explore for oil and to build their own refineries, which would help undercut the monopolistic stranglehold that the international oil companies had on these countries. These activities, which threatened the powerful international oil companies, thus contributed to heating up the Cold War, as the companies frantically and successfully called on their home governments for assistance. The clarion call of the U.S. National Petroleum Council is reflected in its widely circulated 1962 report on the "Soviet oil drive":

> The ultimate goal of the Soviet Bloc is to expand its political control, destroy freedom, and communize the world, and it uses its monopoly of foreign trade to further these objectives. This, in short, is the problem the free world faces when trading with the Soviet Bloc. . . .
>
> Without a doubt, Soviet oil is the most important element in the Soviet politico-economic offensive in the Free World. The communists are using it to procure vital equipment and technology, to create political unrest and spread communism. It is a weapon with which they hope to destroy the private oil industry.
>
> The seriousness of the Soviet economic offensive requires a concerted effort by the leading countries of the Free World to restrict further imports of communist oil and the export of strategic materials to the Soviets. Individual action is insufficient. . . .

That the international oil companies had considerable success in their efforts with their home governments can be seen from the fact that in 1962 NATO embargoed exports of oil pipeline to the Soviet Union in an attempt to reduce Soviet oil exports to Central and Western Europe. Moreover, thanks to pressure from Western governments, the World Bank, and the International Monetary Fund, a number of underdeveloped countries, most notably India, were blocked from accepting favorable barter deals for Soviet crude oil.

There can be little doubt that the aggressive Soviet foreign energy policy in this period was partly motivated by political, or "noneconomic," considerations. It is true that the exports of Soviet crude oil frequently gained the country badly needed foreign exchange and/or useful consumer products. However, the Soviet offers of technical and financial assistance to governments of underdeveloped countries to build up their own oil industries was clearly costly, since it meant providing men and equipment which could be used at home for expanding the country's own oil industry. It only made sense within the overall framework of Soviet foreign policy in this era.

That is, with the country locked in mortal combat with the United States, particularly following the Bandung Conference of 1955, the Soviet Union under Khrushchev undertook a policy of actively wooing the underdeveloped countries. Oil was seen as a very useful lever for gaining a foothold for Soviet influence in the Third World. For example, the Russians unilaterally cancelled a contract to deliver oil to Israel in November 1956 following the Israeli invasion of Egypt. Moreover, insofar as use of this lever would hurt the international oil companies, it had double value, since Soviet theoreticians at that time placed heavy emphasis on the key importance of the oil industry as a source of Western power. Thus it was argued in an authoritative Soviet publication that:

> It should be borne in mind that oil concessions represent, as it were, the foundation of the entire edifice of Western political influence in the [less developed] world, for all military bases and aggressive Blocs. If this foundation cracks, the entire edifice may begin to totter and then come tumbling down.

The shift away from an aggressive Soviet foreign energy policy began in the early 1960s, and continued during the rest of the decade. In this period the Soviet Union became much less active in offering to assist underdeveloped countries to build their own petroleum industries. In my view a crucial factor behind this change was the shift in overall Soviet foreign policy to a much more accommodationist stand toward the United States after coming to the brink of nuclear exchange in the 1962 Cuban missile crisis.

It is understandable that the Soviet government's reaction to the Cuban events would lead to a particularly sharp moderation of Soviet foreign policy in the energy area. One of the most important links in the chain which led to the U.S.-Soviet clash over Cuba was that in 1960 the Cuban government had nationalized the refineries of the international oil companies on the island. The government did this because the refineries refused to process Soviet crude oil which the government had obtained in a barter deal for Cuban sugar; the nationalization in turn had led the U.S. government to cut Cuba off from the U.S. sugar market and ultimately to break off relations with Cuba entirely. Thus, the Soviets had learned again from firsthand experience that threatening the interests of the international oil companies was truly playing with fire.

What shape Soviet foreign energy policy will take in the future is closely tied up with the complex interaction between Soviet supply/demand of energy resources, and broader questions such as the general road of the Soviet economy and overall Soviet foreign policy. In my view, several factors seem likely to be crucial.

First, the drive toward increasing production of consumer durables, and automobiles in particular, will continue and will place increased strains on energy resources. Second, it is unlikely that the Soviet Union will reach such a sufficiently secure accommodation with the West that it will be able to afford to relax its interest in the vast oil resources of the Middle East. This is particularly true since in recent years the Soviet Union has been warning the Eastern European countries that in the future they will have to rely more on the Middle East and less on the Soviet Union for their growing oil imports.

Moreover, the increasing effort of China to play a role in

the Middle East is bound to intensify Soviet involvement in the area, since it apparently now sees China as its main international enemy. Thus, the overall conclusion is that the Soviet Union, like the United States and the rest of the developed countries, has its own vital interest which will preclude a passive foreign energy policy, particularly as regards the Middle East.

Having examined the historical development of Soviet energy policy in some detail, we can more briefly look at China's energy policy over the last twenty-five years, since its main features in many ways parallel that of the Soviet Union before 1960. However, it is first helpful to examine China's energy policy within the historical context of the country's own energy experience.

Petroleum enjoys a unique place in Chinese history in that it was the first Western product that China accepted voluntarily. Until kerosene was imported in the 1870s the Chinese insisted on silver in payment for its silk and tea exports, while opium was forced into the country's bloodstream by the British victory in the Opium War of the 1840s.

Standard Oil dominated the market with exports from its Bayonne refinery and by 1884 Standard's kerosene exports to Asia accounted for one-fourth of its total exports. Royal Dutch Shell was the leading competitor with its oil from the Dutch East Indies and Russia. At first kerosene was used for lamps in business houses and rich merchants' homes, but wide distribution of a thirty-cent oil lamp vastly increased the market for oil. By 1911 kerosene was China's third largest import, after opium and cotton, and consumption continued to grow thereafter, although indigenous oil production was negligible.

On the eve of the Communist victory in 1949, China imported over two million tons of petroleum products annually. With no refineries within the country, China, like most underdeveloped countries, was simply a highly profitable marketing outlet for the major international oil companies. One of the first acts of the new Communist government was to nationalize the existing marketing facilities of Standard Vacuum (Exxon and Mobil), Royal Dutch Shell, and Caltex (Standard of California and Texaco). Since these were among the largest U.S. investments in China, needless to say this contributed to the ferocity of the U.S. government's hostility to the new regime.

Unlike Russia, however, in China the new Communist government inherited an economy where oil was a very minor factor. Though the international oil companies insisted China had no oil, very little exploration had actually been done in the country, and indigenous crude oil production was only about 400,000 tons per year, or 1 percent of the country's total energy supply. Because wood was scarce, its use as a fuel was negligible. Despite vast water resources, the country's hydroelectric power was virtually untapped. Hence, almost by default, coal, which was widely dispersed in the country, was overwhelmingly king. In 1952, after a period of reconstruction, coal production had reached its 1942 peak of sixty-six million tons, and accounted for 98 percent of China's total energy consumption.

In general, the early goals of Communist China's energy policy appear to have been similar to those of the Soviet Union. On the demand side, top priority was placed on electrification of the country and on providing fuel for heavy industry. On the supply side, the emphasis was placed on developing indigenous resources rather than relying on imports. Hence, while widespread exploration for petroleum took place, in the interim before such exploration could pay off, the growth of petroleum consumption was restricted, and the energy sector continued to be overwhelmingly dominated by growing coal production.

Thus, it is estimated that between 1952 and 1962, when total energy consumption increased more than fourfold, coal still made up 95 percent of the energy used, while indigenous oil production, which had increased fifteenfold, still accounted for only 4 percent of energy consumption. Moreover, in the early 1960s China still imported about four million tons of crude oil products annually, mostly from the Soviet Union, while itself refining only six to seven million tons.

The falling out between the Soviet Union and China in 1960, when the Soviet Union abruptly suspended its economic and technical assistance, showed vividly the dangers of relying on a foreign power for supplies of such a vital commodity as petroleum. Thus, it was reported in *World Petroleum* in March 1961 that:

> An acute shortage of petroleum and products is reported from Red China, believed by Western observers to be a result of calculated Soviet economic pressure. Most of China's petroleum is supplied by the Soviet Union. The shortage appears to be growing ever more acute as the scale of Chinese industrialization increases. Strenuous measures are now being taken to meet the situation. They include cuts in public transport services. There has also been noticed a marked reduction recently in the activity of Chinese jet fighters when trying to intercept Nationalist Chinese reconnaissance planes. China's jet and aviation fuel is provided entirely by the Soviet Union.

The wisdom of the previous vast oil exploration effort—in which for example in 1956 more than ten times as much oil drilling was done as had been done in the whole 1907–1948 period—was amply proven by subsequent oil production figures. According to United Nations estimates, Chinese crude oil production doubled between 1961 and 1966, which allowed a 60 percent increase in petroleum product consumption, while petroleum imports dropped by one-third. By today China's crude oil production has reached about fifty million tons, or quadruple the 1966 level, and has allowed the country to become a crude oil exporter. Moreover, it appears likely that in the future China will be a major crude oil producer (it already ranks about fourteenth in the world).

While these statistics are quite impressive in absolute terms, to truly appreciate the tremendous achievement of the Chinese in the energy area, as well as its effects on the life of the average Chinese citizen, one has to see how other underdeveloped countries have fared in this area. Thus, in the next chapter, where we examine the role of the oil-importing underdeveloped countries in the world energy picture, we will conclude with a comparison of the energy performances of India and China. Such a comparison will shed additional light on the performance of market versus planned economies in the Third World, as well as the relationship between energy and the overall society.

7
The Oil-Importing Third World

The oil-importing underdeveloped countries constitute close to a hundred countries which make up the great bulk of the Third World, ranging from Argentina to Zambia. Before World War II the typical oil-importing underdeveloped country was a colony. Since it was poor and had virtually no industry, it was a relatively small consumer of oil. However, even though the international oil companies' sales in these countries were relatively small, they were highly profitable because the companies made very little investment within these countries. Thus, essentially these nations served as highly profitable marketing outlets for the low-cost crude oil which was produced and refined in giant refineries owned by the companies in North America, the Middle East, and the Caribbean.

After World War II most of the underdeveloped countries gained nominal independence and could theoretically choose an energy policy in line with their goals to promote economic development. While there is great diversity among these nations, most of them share certain common energy and petroleum problems, arising from several interrelated vicious circles of underdevelopment in energy and petroleum resources which impede economic development.

First, industrialization and modernization generally require even more rapid growth in energy consumption than in the total output of the economy. Second, many of the countries have a shortage of indigenous energy resources, due either to nature, lack of knowledge as to the resource base, or failure to exploit proven resources. Third, most of the countries are short on foreign exchange which could be used for importing additional needed energy, particularly petroleum. Fourth, many of the countries rely to a lesser or greater degree on foreign aid to help fill their foreign exchange gap. Fifth and finally, their internal petroleum sector, whether well or poorly developed, is generally dominated by the international oil companies whose home governments, particularly the United States, are the primary sources of this foreign aid.

In addition to these common energy problems, the oil-importing underdeveloped countries as a group share the twin general goals of rapid economic development with maximum political independence. However, while all may agree on these general goals, or at least pay lip service to them, there are a wide variety of views as to the best economic and political policies for attaining them. Petroleum policy is both an important component of general policy, and also a major influence on these other elements of policy. Hence in analyzing petroleum decisions, a number of factors have to be considered, especially the following: 1) expected or planned changes in the output structure of the economy as a whole in the course of development; 2) costs of various energy sources; 3) flexibility of different energy sources; 4) locus of ownership and control of energy resources; 5) economic, political, and social repercussions of different energy growth patterns. These five sets of considerations are crucial for any underdeveloped country attempting to establish an energy and petroleum policy consistent with its other goals in the area of overall economic development. Below we sketch out some aspects of these considerations.

The pattern of economic development which has emerged in many underdeveloped countries has been one of relatively rapid industrial growth and relatively stagnant agricultural production. This has profound implications for energy policy. Because energy input per unit of output is generally far higher in industry than in agriculture, this pattern means the energy sector will almost always have to grow far more rapidly than the overall economy.

In addition, the specific composition of the overall economic development plan has a crucial relation to energy and oil policy, both as cause and effect. For example, the problem of transport development—whether to stress rail, road, or river—is intimately bound up with energy resource availability.

The cost of different energy sources is of course a key factor in energy policy. Several dimensions of "cost" should be taken into account. First, what are the social costs (the ultimate costs to the country as a whole) as compared with the private costs (the immediate money costs to the individual or firm)? Second, what are the short-run and long-run costs (the current operating costs and the capital costs)? Third, what are the foreign exchange costs as opposed to the indigenous currency costs?

Another important consideration in energy policy is the "flexibility" of different energy sources. Generally speaking, oil is the most flexible fuel resource available to underdeveloped countries. It can be used directly for heat energy (in industrial boilers, home cooking, etc.), for all forms of transportation, for lighting, and for generating that most important secondary energy source, electricity. Oil can fill specific requirements that cannot be met by other energy sources, at the present stage of technology (e.g., petroleum fuels for road or air transport). Specific petroleum products can also be used for various purposes (e.g., kerosene for home lamps or for aviation fuel, diesel oil for locomotives or stationary engines). And the proportion of various products derived from a barrel of crude oil can be varied somewhat within a refinery.

The locus of ownership and control of energy is a crucial concern of governments the world over. There are three basic dimensions to the problem: 1) physical location—foreign or indigenous; 2) ownership location—foreign or indigenous; 3) social control—private or government. In most underdeveloped countries the decision on foreign versus indigenous ownership generally follows from decisions on the other dimensions of the location problem. For example, a policy decision to rely primarily on indigenous energy resources may effectively rule out foreign ownership because of lack of foreign interest in hydroelectric or coal development. Again, a decision to have government ownership of energy resources automatically rules out foreign ownership.

Intelligent policy on energy must also take into account the important economic repercussions which may result as a "byproduct." For example, development of hydroelectric resources frequently serves nonenergy goals, such as irrigation and flood control. Probably the most universal economic byproduct consideration in formulating petroleum policy is its impact upon the potential for an indigenous petrochemical industry which can serve as an important base for economic development.

Finally, aside from directly economic repercussions from any given petroleum and energy policy, there are often important political and socioeconomic ramifications to be considered. For example, does the policy leave the country vulnerable to an energy boycott, and how important is that to the country? What would be the impact on foreign relations of choosing government ownership of energy resources? Does the country want a nuclear power industry as a base (or cover) for developing nuclear weapons? How will resolution of the public versus private sector question affect the balance of political forces within the country?

The social repercussions of different energy patterns may also be very significant in the long run. For example, undertaking a major hydroelectric project complete with irrigation and flood control may ultimately revolutionize conditions in the local area affected. It may greatly increase productivity, raise the standard of living of local inhabitants, and radically change the social relations among various groups within the area. Again, since most rural areas are not electrified and oil lamps are the only source of light, a decision to allow and promote growth in kerosene consumption may be a necessary adjunct of any campaign to promote literacy. Conversely, blocking growth in kerosene demand may have a negative impact on literacy and may thereby slow down increases in productivity and positive changes in social relations.

The potential ramifications of energy and petroleum policy can thus be seen as widespread, affecting both internal and external relations. At the same time, for just this reason, development of petroleum policy itself is ultimately shaped by the internal and external power relationships.

As a matter of historical fact, despite the theoretically wide range of policy choices, most of the oil-importing underdeveloped countries ultimately built their energy economies around imported oil controlled by the international oil companies. Thus, despite the fact that most of these countries produce relatively little crude oil, since World War II they have become so dependent on oil that it now accounts for the overwhelming bulk of their greatly increased total energy consumption: the underdeveloped countries as a whole derive three-fifths of their energy from petroleum and one-sixth from natural gas.

Several factors account for this situation. First, in many countries coal or hydroelectric power has not been available or fully exploited. Second, even where coal was available, notably in India, the existence of free markets in these countries has allowed the international oil companies to push relatively cheap oil into their economies, at the expense of balance of payments problems and neglected domestic resource development. In the light of the fact that even Europe after World War II could not stand up to the international oil companies, it is hardly surprising that these weaker countries failed to do so.

Moreover, because the companies had low-cost crude oil available from Venezuela and the Middle East, they had no incentive to explore for oil within the oil-importing underdeveloped countries. As a rule, if they evinced any sign of interest in indigenous oil exploration, it was for the purpose of blocking the governments of the countries from undertaking such exploration themselves. In line with this, the international oil companies used the power of their home governments and their influence in world organizations such as the International Monetary Fund and the World Bank to pressure the governments of the oil-importing underdevel-

oped countries not to use public funds to undertake indigenous oil exploration.

The nominal argument used by these "aid-giving" foreign
institutions was that "aid money" should not be diverted to
oil exploration (or building government refineries) because
the international oil companies had the vast pools of capital
and experience needed for this task. The fact that, as these
sophisticated foreign agencies knew so well, the international
oil companies had little incentive to explore for oil despite
their capabilities was ignored. The real aim of these foreign
agencies was to support the international oil companies in
their profit-making activities.

There are a number of reasons for the herculean efforts of
the international oil companies to block oil-importing Third
World countries from developing their own publicly owned
oil sector. For one thing, while many of the individual markets were relatively small, in total these Third World countries accounted for perhaps 10 percent of world oil imports
in the early 1960s. Moreover, because these countries were
generally in the earliest stages of industrialization, and had
been virtually untouched by automobilization, the companies
had high hopes that their growth rate of demand for oil
would be far higher than that of the developed countries; in
other words, the oil company goal of "oil for the lamps of
China" had now been generalized to the Third World as a
whole.

Further, because of the general ignorance within these
countries about the international oil industry, as well as the
countries' relative weakness, they have been easy victims of
superexploitation by the international oil companies. For example, with regard to the prices charged, a leading oil expert,
P. H. Frankel, has noted that in the 1950s the oil companies'

relationship to the miscellaneous oil-consuming countries outside
of Western Europe was still in the Dark Ages, as it were . . . there
was no competition at all except for that between the few established major oil companies, and this, commerce being what it is,

resulted in the maintenance of price levels and profit margins which were considerably higher than they were elsewhere—even after allowing for the substantial cost of transport and handling incurred, coupled with the comparatively small turnover. It was said, not without some justification, that the poorer a country was, the higher were the prices it paid for its oil.

Again, a recent report of the U.N. Secretary-General on petroleum needs in the oil-importing developing countries states that even now "the great majority of the . . . countries in the developing world appear to be paying higher than competitive prices." (It should be noted that because the oil revenues of the governments in oil-exporting underdeveloped countries are based on fixed reference prices, they do not derive any benefit from this overcharging.)

Finally, the oil-importing underdeveloped countries are of significance to the international oil companies because as a group they occupy most of the world's land mass and have the majority of the world's offshore waters where oil is increasingly found, and hence are important future potential sources of oil. Since most of these countries have had little if any serious exploration for oil, owing to the companies' restraining policy discussed previously, who can tell which of them might in the future turn out to be an important source of oil or of other energy resources? After all, it was not so long ago that major oil exporters like Libya, Algeria, Nigeria, and Ecuador were themselves oil-importing countries. Thus, particularly now that the OPEC countries have become less reliable sources of crude oil for the international energy companies, it is important for the latter to have access to any potential energy resources.

The position of the international oil companies in the oil-importing underdeveloped countries after World War II has been complicated by the changes from colonial to independent status. It is true that while many of these countries paid lip service to socialism, they were essentially capitalist market economies. However, the existence of an organized Left and mass pressure in some countries, and nationalist military/ bureaucratic elites in others, has led to the growth of a substantial public sector in the oil area. The profit-maximization goals of the companies generally lead them to oppose these efforts on the part of the underdeveloped countries. The companies fear that government oil activity in any country may interfere with their balancing of their worldwide interests, or reduce their profits within the country, or ultimately lead to nationalization.

The broad aims of the international oil companies vis-à-vis the oil-importing underdeveloped countries are specifically manifested in three areas: oil exploration and production, refining, and transportation. With regard to exploration, as noted before, since the companies generally had large quantities of low-cost crude oil outside the oil-importing countries, they had little incentive to explore for oil within the countries and a great incentive to keep the governments from undertaking exploration. In the area of refining, the companies' main interest was in obtaining outlets for their external crude oil; hence they generally sought to prevent government ownership of refineries, particularly since this might have led logically to greater governmental efforts in the field of oil exploration. As for transportation, the companies wanted to maintain control over the international tanker fleet so they could continue to charge the highest profit-maximizing price for shipment.

Given this setting, the crucial policy issue for the underdeveloped countries seems to be the extent to which they should rely on the international oil companies, as opposed to developing an indigenous, and normally public, oil sector. By and large, the international oil companies and their supporters have been quite successful in influencing these policy decisions in a direction favoring private foreign investment. The question of foreign or government ownership of refineries in the Third World is one crucial example. Refinery ownership is a key issue because government ownership of a refinery, which is a relatively riskless investment, can help to insure that the country is not overcharged for crude oil, as has so often been true when the refineries were owned by integrated companies which also supplied the oil for refining. And, of course, since petroleum products are so crucial for defense purposes, it is clear that refineries are truly a "commanding height" of the economy.

Nevertheless, despite the obvious value of government ownership of the refining sector, in the oil-importing Third World countries as a whole, governments own outright less than half of the total refining capacity. The private foreign sector owns about three-eighths, and about one-sixth is owned by mixed ventures, mostly equal partnerships between governments and the international oil companies. Moreover, the aggregate figures actually obscure the relative lack of governmental participation in the Third World refining industry because they are heavily weighted by the data for Latin America, with its long history of state ownership in the oil sector. Thus, while on a worldwide basis the ratio of government-owned refining capacity to privately owned capacity is 55:45, in Latin America it is 70:30, while in Asia it is 50:50 and in Africa only 20:80. Looked at another way, in half of the oil-importing underdeveloped countries all refineries were completely owned by the international oil companies, while in about 30 percent of the countries, government and the private sector shared ownership; in only one-fifth of the countries was the government the sole refinery owner.

Similarly, the international oil companies have been largely successful in blocking or slowing oil exploration in the oil-importing underdeveloped countries, thereby retaining lucrative markets for their low-cost Middle Eastern, African, and Caribbean crude oil. Thus, Chase Manhattan Bank data reveal that in the 1962–1972 period, when a total of $83 billion was spent for capital expenditures on oil exploration, development, and production in the non-Communist world as a whole, only $14 billion was spent in the Third World, excluding Venezuela and the Middle East. Of this total the great bulk was undoubtedly devoted to expenditures in other Third World oil-exporting countries such as Indonesia, Nigeria, Algeria, and Libya; moreover, only $3 billion was used for the crucial expenditures on gathering geological and geophysical data.

Nevertheless, even taking the $14 billion figure as an upper limit, this amounts to only one-sixth of total capital expenditure for oil exploration, development, and production, compared to a figure for the United States of more than three-fifths of the world's total. At the same time, the oil-importing Third World countries account for two-thirds of the non-Communist world's land mass, while the United States accounts for less than one-tenth. Considering that the United States has already been intensively explored, while much of the area of the Third World has never been explored, it is clear that these figures reflect the international oil companies' distinct lack of interest in seeking oil in the Third World. The extent to which this will change as a result of the 1973 shake-up of the oil industry is difficult to predict at this juncture.

The cost to the Third World of this kind of energy policy, with reliance on foreign ownership and imported energy resources, has been very great. A classic and important case study is that of India, which contrasts sharply with China, where the government has vigorously developed its own indigenous energy resources. In the indigenous crude oil exploration area, a limited Indian effort has succeeded in raising production from only a negligible amount in 1952 to about eight million tons per year today, and the Indians have been reduced to bringing in foreign contractors to explore on a production-sharing basis. In China, on the other hand, during the same period production increased from the same low level as India to fifty million tons per annum today, or six times as much. Similarly, in the last twenty years India has only doubled its coal production, while China has increased it more than sixfold, to a level more than five times as great as India. As a result, while in 1952 China produced about 50 percent more total energy than India, and was about even on a per capita basis, today China produces 300 percent more total energy than India. It is a sad commentary on the Indian performance that the only energy statistic in which it leads is the growth of oil imports, which have increased from three million tons in 1952 to about sixteen million tons today;

China has gone from importing about two million tons annually in the earlier period to being an oil exporter today.

Not surprisingly, a country's performance in the energy sector is reflected in the performance of other vital sectors of industry. One of the most important of these is iron and steel, a key to overall industrialization, for which energy resources are crucial inputs. Indian production of pig iron has increased from 1.7 million tons in 1952 to about eight million tons today (little more than was produced in 1965), or a four-fold increase. Chinese production, by contrast, has catapulted from 1.9 million tons in 1952 to 27 million tons in 1971, or a fourteen-fold jump.

It should also be noted that India's relatively slow industrialization has been associated with a tremendous increase in the country's external debt, from virtually none in 1952 to over $5 billion today (excluding the enormous food loans from the United States, most of which were finally written off because there was no chance India could repay them). This growing debt, in turn, is associated with increasing incursions on the country's sovereignty. For example, as noted earlier, in the early 1960s the Indian government was essentially blocked from accepting a favorable Soviet barter offer for crude oil because of its weak position vis-à-vis the West. Again, in the 1962 war with China, the Indian government had to bring great pressure to bear merely to get the private refineries to produce jet fuel instead of the kerosene which was much more profitable for the companies. Later, in the mid-1960s when the Indian government sought to promote a public fertilizer sector at the expense of private foreign investment, a combination of poor crops in India and long-standing debt relations to the West made the Indian government quite vulnerable, as the *New York Times* notes in 1966:

> Much of what is happening now is the result of steady pressure from the United States and the International Bank for Reconstruction and Development, which for the last year have been urging a substantial freeing of the Indian economy and a greater scope for private enterprise.

The United States pressure, in particular, has been highly effective here because the United States provides by far the largest part of the foreign exchange needed to finance India's development and keep the wheels of industry turning.

Call them "strings," call them "conditions," or whatever one likes, India has little choice now but to agree to many of the terms that the United States, through the World Bank, is putting on its aid. For India simply has nowhere else to turn.

All this is in striking contrast with China. While it received significant amounts of Soviet aid in the 1950s, much of the value of this was reduced by the sharp blow to the economy caused by the sudden halt of Soviet assistance in 1960. Nevertheless, China repaid all of its debt to the Soviet Union by 1965, and now has relatively little debt to other countries.

To a great extent the dangers and inherent weaknesses of energy policies which rely on foreign-owned energy resources was masked for countries like India by the declining oil prices in the 1960s. These declining prices helped to offset the cost from rising volumes of imported crude oil, and hence did not make for an unmanageable burden on the countries' economies. Now, however, with the big increases in oil prices since the early 1970s, particularly following the October war, the chickens have really come home to roost.

While this will be discussed in more detail in the next chapter, in general it should be noted that for oil-importing Third World countries like India the tremendous jump in oil costs has left the balance of payments a shambles, badly hit fertilizer consumption, and drastically threatened industrial growth, if not the very lives of many of the people in the Third World. With economies still dominated by market forces, and governments relatively impotent to take action in the face of them, the energy crisis has been compounded into a disaster of capitalist depression and stagnation. Thus, in a perceptive article on "The Agony of India: The Country Is Beset by Multitude of Woes," the *Wall Street Journal* noted:

The oil crunch plus disrupted coal deliveries plus scattered shortages of hydroelectric power (because of low water levels in some reservoirs and equipment breakdowns) has resulted in a general, nationwide energy shortage. Demand for power meanwhile is rising. Indian industry is also faced with higher prices for imported capital goods and is faced with reduced consumer demand because of inflation. To all of this must be added transport problems, strikes, and the traditional difficulties of doing business amid the controls and red tape of Indian bureaucracy. The unsurprising end result is stagnant industrial productivity. Many factories are operating far below capacity, unemployment is increasing, shortages of manufactured goods are spurring inflation, and export opportunities tragically are being missed.

The situation provides all sorts of painful paradoxes. Fertilizer plants are operating well below capacity while India must shell out foreign exchange for high-priced fertilizer. The steel industry is operating at under 50% of capacity while steel is being imported.

Vicious circles abound. Steel plants aren't getting adequate coal supplies. The coal mines complain that they cannot get the railways to move the coal to the steel plants. The railways complain that they cannot get enough railway wagons to carry the coal. The plants manufacturing railway wagons complain that they cannot get steel with which to build the wagons.

What is the moral of the Indian situation? In my view it is that an energy policy reflecting capitalist market economics is inevitably a disaster for Third World countries. In the long run the only solution to the problem of the energy crisis, as well as the other economic and social crises, is genuine social revolutions in the Third World countries. However, in the short run, before such radical changes can be undertaken, there is a real need to work out some relief for many of these countries in order to save millions from starvation. To see the possibilities for such relief, we must first examine within a global context recent events in the oil industry. This task we undertake in the next chapter.

8
Recent Events and Future Prospects

Recent events have drastically changed the equilibrium of the international oil industry, with major repercussions for all countries of the world. The industry is still in a state of flux, and the new equilibrium is far from clear at this point.

To understand the present situation, it is necessary to review at least the recent historical background. Up until the beginning of the 1970s the Big Seven (five American and two British-Dutch international oil companies) dominated the industry, through their ownership of the great majority of the world's low-cost oil as well as their vertical integration into the refining, marketing, and transporting of oil. The economic power of the Big Seven was backed in turn by the power of their home governments, of which the Anglo-American overthrow of the Mossadegh government in Iran in the early 1950s was one single visible effect. In brief, the enormous profits of the Big Seven were built on their ownership of the world's low-cost crude oil, particularly in the Middle East, combined with the political and military power of their home governments which insured the companies' retention of these crude oil reserves.

In the 1960s, however, the combination of the enormous potential oversupply of cheap oil from the Middle East (with costs at about ten cents per barrel) and the competition from profit-maximizing newcomers to the international oil industry led to a steady decline in the market price of crude oil. Taking Arabian Light as a typical crude, prices in arms-length transactions dropped from somewhere around $2.00 per barrel at the beginning of the decade to $1.25 per barrel at the end. During this period, the Organization of Petroleum Exporting Countries (OPEC), which was born in 1960 in response to the companies' cuts in the posted prices of crude oil, was a relatively weak organization, as we have seen. (The significance of the posted prices was, and is, that producing-country taxes on the crude oil are calculated as a fixed percentage—then about 50 percent—of the posted price, regardless of the market price; thus, in 1960 if posted prices had been cut by twenty cents per barrel, the governments would have gotten ten cents a barrel less from the companies, whose profits would have been increased by that same ten cents per barrel.) However, OPEC generated sufficient pressure to stop the oil companies from cutting the posted prices any further, so that government revenues per barrel remained constant and the declines in market prices were reflected in declining per barrel profits for the companies.

Nevertheless, the per barrel drops in company profits were more than offset by increases in the volume of production, so that company profits on total crude oil sales rose over the 1960s, even if at a much slower rate than OPEC government revenues. Thus, while the gross production of the Big Seven in the Eastern Hemisphere (primarily the Middle East) increased by 157 percent in the 1960–1969 period, their net earnings from Eastern Hemisphere operations increased by only 62 percent. On the other hand, revenues of the Big Four OPEC members from the Middle East (Saudi Arabia, Iran, Iraq, and Kuwait) increased in the same period from $1.3 billion to $3.1 billion, or by 144 percent.

A plausible "static" economic projection at the end of the 1960s would have been more of the same for the 1970s. The most articulate proponent of the steadily falling crude oil price projection was Professor M. A. Adelman of the Massachusetts Institute of Technology. He foresaw that in the 1970s the price of crude oils like Arabian Light would continue to decline from $1.25 per barrel toward a limit involving no profit per barrel for the companies, around $1.00 per barrel (equal to OPEC government taxes plus production costs). Moreover, even that minimum price might be lowered if the OPEC governments started competing with each other by cutting taxes in order to increase sales.

The core of this Adelman projection was the assumption that "competition will out" and that there was no way for monopolistic control to be asserted over the supply of crude oil in order to reverse the fall in market prices. Thus, it was assumed that the decline of the Big Seven's monopolistic power in the 1960s could not be reversed by monopolistic power on the part of the OPEC governments. The reasons for this belief were that the governments were too diverse and disunited by their short-run economic self-interest to agree on a common front against the companies or against the consumers. (That is, Adelman believed, even if all the governments nationalized their oil industries, they would compete against each other because with crude oil costing only ten cents to produce the potential profits from each additional barrel at any price even in the $1.00 range were so great that the governments would inevitably cheat on any price-fixing agreement; in other words, Adelman, as a "Chicago School" economist, believed that in the long run or even in the medium run cartels will not work because the greed of the individual members leads to cheating which brings about the *de facto* if not *de jure* breakdown of the cartel.)

Now, as the reader is quite aware, the 1970s projections based on these assumptions rank with Herbert Hoover's "Prosperity is around the corner" as among the worst in history. Rather than continuing to decline, crude oil prices have jumped as much as tenfold since 1969. Exactly why this dramatic reversal took place in the 1970s is not fully clear even now, but with the benefit of hindsight we can trace some of the forces which led to the change. In the following I will simply present my own interpretation, without attempting a definitive proof.

An important underlying force was the growing weakness of American imperialism in the mid-1960s and after, culminating in its defeat in Vietnam, which was already apparent and irreversible by the end of the 1960s. The whole Vietnam experience (which had gravely weakened the American economy through inflation and balance of payments problems) had also generated a strong antiwar, anti-interventionist mood in the American people. This made it increasingly unlikely in the late 1960s that the U.S. government, the strongest ally of the Big Seven, could intervene physically in the Middle East as it had before.

A second underlying factor was the smashing Israeli victory in the June 1967 war. In hindsight it can be seen that this abject defeat of the Arabs, combined with continued U.S. support for Israel, made it very difficult for any Arab government, no matter how reactionary, to fail to use the ultimate trump card of the "oil weapon" when war broke out again, as it did in October 1973.

On a more immediate level, one of the most significant changes seems to have been the overthrow of King Idris in Libya and the taking of power by Colonel Muammar el-Qaddafi in September 1969. Today, it is hard to realize that in the period before Qaddafi took power the OPEC governments used to bargain with the international oil companies for months and years in order to gain additional crumbs of the potentially enormous oil pie; for example, in 1968 and 1969 Algeria and Libya each had prolonged negotiations with the oil companies over these countries' attempts to achieve an additional ten to twenty cents per barrel in revenue.

Because Libya was a rapidly growing source of ever more important low sulfur crude oil located close to Europe, and a country with a very small population (two million), it had great potential bargaining power vis-à-vis the international oil companies. In the hands of Qaddafi, who is a fiercely anti-imperialist (and anti-Communist), religiously ascetic Arab nationalist, this bargaining position was used to whipsaw the international oil companies. In turn, the achievements of Libya virtually forced the rest of OPEC to use them as a standard for the Teheran Agreement of February 1971 which raised posted prices sharply; Libyan prices were further raised in the Tripoli Agreement of March 1971.

Another factor in Qaddafi's success was that he adopted a policy of ordering the companies to cut production if they would not agree to his demands. This was particularly effective in Libya where newcomer oil companies like Occidental Petroleum were heavily dependent upon Libyan production for their total company profits, and hence were in a much weaker bargaining position than the Big Seven, which had vast worldwide oil supplies.

While due credit should be given to Libya (as well as Algeria, with which it worked closely) for these aggressive tactics, there are strong indications that the "fight" at Teheran was fixed from the start. In the words of Taki Rifai, former adviser to the Libyan Ministry of Petroleum, who, by his own account, was involved fully in the historic 1970-1971 negotiations:

> The attitude of oil companies vis-à-vis claims for higher prices changed strangely during the crisis period. In the early Libyan negotiations in January-February 1970, a top executive of a leading major oil company operating in Libya stated that since his company was forced to accept a price increase, all it could afford would be about a 5¢/Bbl increase, beyond which the company would incur losses in its Libyan operations. A few months later, the same major company spontaneously announced unilateral price increases of much greater magnitude, not only in Libya but also at the Eastern Mediterranean, where it was not subject to any specific claims.
>
> On the other hand, the front of oil companies represented by the "January 16 Group" [1971] did not show any significant resistance to OPEC claims, and the Teheran negotiations almost seemed to be "club discussions" for drawing up the details of a formal agreement rather than to challenge its basic components. The oil companies were there to sign, not to fight.

Moreover, according to the testimony of an oil company insider, the critical turning point in the early Libyan negotiations was the refusal of the major international oil companies, particularly Exxon, to help Occidental Petroleum resist Libya's demands. Thus, Occidental had requested the major internationals to agree to provide it with crude oil from their other sources in the event it found its Libyan supplies reduced, but the majors refused. This predictably forced Occidental to cave in to Libya's demands for increased prices and taxes, which demands predictably the majors would then have to meet.

Surely, if the majors really wanted to keep oil prices down they could easily have provided Occidental with some of their low-cost crude from other countries; their failure to do so suggests one or both of the following. First, that they wanted prices to rise, which would happen if Occidental agreed to Libya's demands in order to save itself from nationalization. Second, that they wanted to eliminate Occidental as a strong competitor, which would happen if Occidental refused the Libyan demands and was nationalized or had its production cut back. In either case, the companies could only gain by refusing support to Occidental.

There is additional evidence that the international majors had the interrelated goals of increasing foreign crude oil prices and eliminating competition. For one thing, as we have noted earlier, raising the prices on foreign crude oil would not only increase the profitability of the majors' U.S. oil, but also the profitability of their other energy sources such as natural gas, coal, and uranium. In particular, since it was already foreseen around this time that the United States in the 1970s was to be far more dependent on crude oil imports than in the 1960s, there was increasing pressure to end the United States oil import quota law. This law, which restricted imports of foreign oil, had kept the price of domestic crude oil more than $1.00 a barrel above that of foreign crude oil.

In fact, in February 1970 President Nixon's own task force studying the law had recommended its abolition and replacement by a steadily declining tariff. This new system would sharply reduce the majors' profitability on both foreign and domestic crude oil. While Nixon, who has always been favorably disposed to Big Oil (and vice-versa, as evidenced by oil industry contributions of over $5 million for his 1972 campaign), rejected his commission's recommendation, it must nevertheless have been clear to the majors that there was considerable danger in that direction. If the price of foreign crude were increased to the level of U.S. crude, then the oil import quota law could be safely abolished without threaten-

ing the majors' profits. Thanks to the rises in foreign crude
oil prices triggered by the early 1970–1971 negotiations, this
was accomplished six months before the 1973 October War.

The necessary piece to complete the picture of a revital-
ized cartel is the U.S. Justice Department's removal in January 1971 of antitrust restrictions on the American companies negotiating at Teheran. According to U.S. government
officials, this was done to give the American oil companies
the ability to present a common front at the negotiations.
This was purportedly needed because of the administration's
concern with "national security" and its fear that, if har-
monious agreement was not reached, the United States might
lose access to Middle Eastern oil.

Whatever the U.S. government's motivation, the removal
of antitrust restrictions on the U.S. oil companies opened the
door for them to make comprehensive plans for worldwide
cartelization. That the companies took the opportunity to
drive a truck through this opening appears to be confirmed
by a top Justice Department official. In announcing in mid-
1974 the end of the antitrust clearance, an assistant Attorney General gave as reasons that the coordinating body of the
companies (the London Policy Group) which "was to be an
ad hoc organization had become a quasi-permanent insti-
tution for oil company cooperation." He noted: "Studies
within it tend to approach sensitive competitive areas of supply, cost, demand, control of downstream distribution and
possible exclusion of independents by means of exclusive
buying-back arrangements."

At the same time, some of the fruits of such efforts were
being uncovered by an investigative body representing the
nation's 5,000 prosecuting attorneys, which found that the
major oil companies were pursuing "anti-competitive prac-
tices and tactics":

> The group, a committee of the National District Attorneys
> Association, said that a five-month probe had uncovered enough
> preliminary indications to warrant formal antitrust investigations
> and prosecutions by local district attorneys across the
> country. . . .

Not only are the oil companies showing monopolistic tendencies in petroleum production and sales, the prosecutors charged, but they are also attempting to gain control over the entire energy industry.

The U.S. government's role in the 1971 negotiations is also quite revealing, in that it reportedly undercut the nominal common front of the oil companies vis-à-vis the producing countries. President Nixon sent Undersecretary of State John Irwin to Teheran in mid-January of 1971, ostensibly to show the U.S. government's support for a common effort by the oil companies to resist higher taxes and prices. In fact, Irwin undercut this effort by advising that the talks be divided into separate discussions with the Gulf countries and Libya and Algeria, thereby leaving the companies open to leapfrog tactics by the OPEC countries.

While many people, such as Professor Adelman and Senator Frank Church, have attributed this action of the State Department, which contributed to the future rapid upward spiral of oil prices, to "bungling," another interpretation is quite possible. For one thing, as we have seen previously, the U.S. government and the State Department in particular have been close to and strongly supportive of the major U.S. international oil companies; as Jack Anderson stated in 1967, "The State Department has often taken its policies right out of the executive suites of the oil companies." Since the major oil companies wanted higher crude oil prices, the State Department's action could be just one more in a long history of service to the companies. Moreover, rising crude oil prices, along with increasing oil company profits, could well be seen by the American government as helpful to the U.S. economy as a whole. This would be true because the increased cost of importing oil would reduce Western Europe's and Japan's balance of payment surpluses, while the U.S. balance would benefit from the increased oil company profits.

The State Department's continuing close relationship to the major American oil companies was also shown by its efforts in connection with Libya's nationalization of Standard Oil of California and Texaco's oil fields in September 1973. The State Department pressured independent U.S. oil companies not to buy the "hot oil" from the Libyan government even though it was economically attractive to the independents. According to the testimony of an executive of New England Petroleum Company, within one hour he received consecutive telephone calls from an official of Texaco, of Standard Oil of California, and of the U.S. State Department warning him not to buy Libyan oil; interestingly, "they all said the same thing. They even used the same words."

In any event, whatever the causes, the Teheran and Tripoli agreements of 1971 and the "participation agreements" of 1972 seem to mark a new era of monopolistic control of oil supplies, this time shared by the companies and the OPEC governments. From these agreements the OPEC governments got a sharp increase in posted prices and in their per barrel revenues, as well as a minority share of the oil companies' production, with the promise of a majority share by the early 1980s. At the same time, however, with the previous oversupply of crude oil now under control, the market prices of crude oil rose even more rapidly than the posted prices and taxes, and the per barrel profits of the companies soared.

From a market low of perhaps $1.25 per barrel in 1969—of which about 10 cents was cost, 95 cents government taxes, and 20 cents company profits—by the middle of 1973 the market price had risen to about $2.50 per barrel, with $1.50 for the government and 80 cents for the companies. Both parties had gained significantly here, but while the government's per barrel revenues had increased by about three-fifths, the companies' had quadrupled.

In addition, the countries had received the right to buy participation in the oil production end. While this was a relatively good deal for the countries, it was not all bad for the companies since it also reduced the pressure for total nationalization. At the same time, "buy back" provisions of the oil agreements allowed the companies to go on selling most of the oil produced in the countries. This was extremely valu-

able in a world of increasing artificially created scarcity.
The fact that the governments were still largely tax collectors—since they sold back most of the "participation oil" and
therefore did not get into the refining and marketing areas—
made them much less of a threat to the operations of the
major companies.

All this of course was but prelude to the vast changes in
the wake of the 1973 October War. The cutback in Arab oil
production, as well as the embargo on sales to the United
States, created near panic buying which sent market prices of
crude oil skyrocketing. In this context the OPEC countries
attempted unilaterally to fix the division of crude oil revenues between the governments and the companies at an
84:16 ratio in their own favor. Thus, in two stages, the
posted price of Arabian Light was raised from $3.00 per
barrel before the war to over $11.00 per barrel by December
1973. The government's per barrel revenues from this Arabian Light thereby leaped from less than $2.00 per barrel to
over $7.00 per barrel.

This attempt by the OPEC governments to set the level of
crude oil profits was undoubtedly a blow to the big oil companies. More direct blows were also sustained from the Iraqi
nationalization of all American and Dutch oil interests. In
addition, the new climate destroyed the timetable for gradual
transfer of majority ownership of the oil fields. Instead of 51
percent for the governments by 1982, it appears that something like 60 percent in 1974, possibly escalating in the future (the deal negotiated by Kuwait in 1974), will be the
minimum goal of the OPEC governments.

On the other hand, the developments stemming from the
October War have also had their bright side for the international oil companies, and particularly the Big Seven. Thus,
even the original OPEC target, now abandoned, of an 84:16
split on profits still would have left a profit for the companies on Arabian Light of $1.20 per barrel, or 50 percent
more than the per barrel profit level before the October War.
Moreover, the OPEC split was based on setting posted prices
at 40 percent above what it considered market prices. But,

since most crude oil is sold in integrated channels within the international companies, there was no guarantee that the companies were not selling their crude oil at prices far above OPEC's estimates of what market prices would be in arms-length deals. Certainly with crude oil being auctioned after the October War at prices in the $15-$20.00 per barrel range, it is highly unlikely that the international oil companies were selling crude at OPEC's estimated $8.00 per barrel level. Hence, in reality the companies were able to use their control of market prices to increase their crude oil profits well above OPEC's target limit.

In addition, the real impact of the speeded-up movement for majority participation and ultimate nationalization is far from clear. As long as the OPEC governments combine these steps with agreements to sell back most of the crude oil to the companies for them to market, then the effect may be more one of form than substance. What the companies are primarily interested in is the quantities of oil which they can draw and the per barrel profits on each, and if satisfactory arrangements can be made for them on these questions, they are quite willing to accede on formal questions of legal title. Witness the much ballyhooed "nationalization" of the oil industry by the Iranian Shah in 1954, which was the figleaf which imperialism installed in place of the real nationalization by Mossadegh.

The real danger to the international oil companies is nationalization in which the government of the oil-producing country takes command of the crude oil supplies and either uses them internally, by building indigenous refineries, or markets them directly to foreign buyers. This would mean that the oil-exporting countries could deal directly with foreign governments or refiners or marketers in consuming countries, thereby cutting off the oil companies' crude oil and gas profits, and ultimately their refining profits. Little wonder then that the September 1973 Libyan nationalization discussed above brought the State Department into quick action on behalf of the majors, just as fear of real nationalization in the leading oil-exporting countries has pushed the major oil companies to seek concessions for oil and gas exploration all over the world.

Returning to the brighter side for the companies, the rise in OPEC oil prices has greatly increased the companies' profitability on various energy sources in the United States. Since the companies were making some profit on their domestic crude oil when it was selling at $3.00 per barrel, they are obviously profiting fantastically now that it is up to $7.00 per barrel and more. It has been less widely noted that coal and uranium prices have also shot up drastically, and what is more, bids to supply coal and uranium at today's prices have virtually dried up, indicating that the energy companies (read the big oil companies) expect even higher prices in the future.

Finally, the October energy crisis has greatly weakened the American environmental movement in the energy area, leading to the pushing through of the Alaskan pipeline and extended offshore drilling for oil, as well as a speeded up shale oil program. (In a related area the lid has also been partially lifted on natural gas prices.)

The upshot of all these events has been a very sharp rise in the big oil companies' profits, for now and in the foreseeable near future. Thus, in 1973 Exxon's profits were up 59 percent, the Big Five American companies up by over 50 percent, and the same for the leading thirty-one companies in the world. The forecast for the period ahead is more of the same, with profits from domestic operations expected to rise "spectacularly" while foreign profits rise "moderately." One study has forecast that the pretax cash flow to oil companies from U.S. operations in 1974 will jump by $16 billion (an increase which is more than three times the level of their 1973 profits from U.S. operations).

Who are the losers from these changes in the international oil industry? Insofar as the price of crude oil has shot up tremendously, all oil-importing countries have been hurt, but obviously some have been hurt more than others. In human terms probably the most badly hurt will be the great majority of underdeveloped countries which rely on imported crude

oil for their main energy source. This is because their economies are always in such weak condition that any adverse
event can easily topple them; as one Third World representative discussing the impact of the energy crisis on these countries put it, "In an anthill the morning dew is a storm."

Thus, the World Bank has lamented that oil price increases,
which have jumped Third World oil imports from $2 billion
in 1969 to $15-20 billion in 1974, have completely wiped
out all the effects of foreign aid given in recent years. If the
Bank is not simply shedding crocodile tears over this event,
then it should hide its head in shame, since its own policy of
not only refusing to loan money to governments of underdeveloped countries for use in developing their oil sectors,
but actively opposing such activities, has contributed heavily
to the drastic plight these countries are now in. (China, on
the other hand, is a good example of what large-scale use of
public resources for oil exploration can accomplish, as I have
shown.)

In any event, for a country like India, the increased foreign
exchange bill for crude oil imports, to well over $1 billion in
1974, equals more than one-third of its total exports, and
undoubtedly means that imports of other vital commodities,
including foodstuffs and fertilizers, will have to be cut. The
result, in the poor underdeveloped countries, then, is not
adequately measured by percentage changes in GNP, but is
truly registered in bellies bloated from malnutrition and increased death.

Given this, it is most heartening that the OPEC countries
have expressed their willingness to provide financial assistance to the oil-importing Third World countries to help al-
leviate this situation. Moreover, an OPEC country, Algeria,
took the lead in proposing the historic April 1974 United
Nations special session on raw materials in developing nations. At this General Assembly meeting an oil loan fund was
established for the poorest countries, and support provided
for all Third World countries to raise prices on their own raw
materials so as to help change the historically unfair terms of

trade between underdeveloped and developed countries.

The second group of losers are the developed countries of the world, but the relative extent of their losses depends in part on the proportion of their energy supplies made up of imported oil. In this regard Japan is worst off, with Western Europe in the next worst position, while the United States is much better off, particularly because most of the profits of the international oil companies return to the United States. In fact, it has frequently been suggested that the United States has not been averse to certain aspects of the energy crisis, since it has weakened Western Europe and Japan to its own benefit—witness the rapid strengthening of the dollar vis-à-vis other currencies following the October War. This is a point to which we shall return later.

Even within the United States, however, there are winners and losers from the energy crisis. The major winners, of course, have been the oil companies, and to a secondary extent the defense industries which have been strengthened by the increasing world tension. Major losers have been the giant auto industry and its supplying satellites, the electric and gas utilities, as well as the widespread leisure industries dependent on the automobile and travel, such as motels, resorts, boating, summer homes, etc. One of the things which is likely to come out of this "energy crisis" is a sharpening of tensions between capital invested in different industries of the economy, particularly insofar as there is suspicion that the crisis has been artificially created by the oil companies and/or that they are not doing and will not do enough to overcome it in the future.

Thus, the recent demands of many congressmen for a more active government role in the oil industry, ranging from gathering data on energy resources and prices all the way to establishing a government energy corporation, reflect not only pressure from ordinary constituents. They also must be responding to demands from leading sectors of big business that the state move to help restore equilibrium among various industries of the economy.

While the amount of capital invested in the auto and related industries and the utilities is obviously enormous, the oil-military complex has up until now had the dominant say in American government circles. Undoubtedly the struggle between the large sectors of industry which have been hurt by the energy crisis and the oil-military complex will not end quickly. Thus we have the recent lawsuits which have been brought by various utility companies charging the oil companies with illegal restraint of trade which has increased the cost of fuel supplies. In the long run, however, what is most likely is that within the United States there will be compromises worked out such that at least American industry as a whole gets adequate energy supplies—that is, a compromise somewhat like that shaped between the international oil companies and the domestic producers in terms of the oil import quota law which ruled for many years—even if at much higher prices.

Given this general background, we can now turn to the critical question: whither the international oil industry? Because of the enormous size and influence of this industry, this question is largely synonymous with "whither the international economy?"

To a great extent the answer to these questions hinges upon an analysis of what the vast increases in OPEC government revenues would mean to Western balance of payments and monetary problems. The crucial problem of OPEC government revenues arises because they have risen from $6 billion in 1969 to $23 billion in 1973 and at current oil prices are projected to increase to close to $100 billion in 1974. Saudi Arabia's revenues alone are projected to rise from less than $1 billion in 1969 to $20-25 billion in 1974. The problem thus reduces to two related questions: how can the importing world continue to pay for the oil it needs, and what might be done with the enormous monetary reserves that the OPEC countries could pile up—at the 1974 level of revenues, OPEC countries could take in well over $500 billion by 1980, a sum greater than total world foreign exchange reserves.

Without drastic modifications, the future situation seems

untenable for Japan and Western Europe. The latter's oil imports, which amounted to $10 billion in 1969 and $20 billion in 1973, are expected to jump to $50-60 billion in 1974. This means Western Europe's oil imports already amounted to 10 percent of its total export revenues in 1973 and could reach 20-25 percent in 1974. For Japan the situation could be even worse. Oil imports rose from $2 billion in 1969 to $7 billion in 1973, at which point they already amounted to over 20 percent of total exports, and in 1974 the proportion could reach one-third. Clearly, for Western Europe and Japan as a whole such levels of oil imports are not sustainable for very long. Moreover, the gravity of the situation is compounded by the fact that within Western Europe some countries, notably Italy, were in serious financial straits even prior to the recent oil changes.

The United States would also tend to be hard hit in the balance of payments area because not only have prices jumped, but U.S. imports are rising much more rapidly than those of Western Europe or Japan as U.S. production levels off. Thus, U.S. oil imports have risen from $3 billion in 1969 to $8 billion in 1973, and an estimated $25 billion in 1974, which would amount to about one-fourth of total exports.

However, the key to the better position of the United States is the enormously growing foreign profits of its international oil companies. While current data are not readily available, even if the oil companies were making only $1.00-$1.25 barrel on OPEC crude oil, this would amount to about $10 billion. Moreover, many more billions of dollars are clearly being made from refining and marketing abroad, since foreign product prices have also shot up greatly. With these profit inflows helping to offset the cost of America's oil imports, the share of total exports needed to pay for vital oil supplies for the United States would tend to be relatively low compared to those of Western Europe and Japan. And when one takes into account the fact that much of Western European trade is within the bloc, so that its net exports to the outside world are far lower, then it is even more likely that the changed oil situation would be a relatively greater burden on Western Europe. Furthermore, the United States has more room to maneuver in that it can increase production substantially within its borders by producing from formerly unprofitable shut-in wells, in addition to expanding quick-return development and exploration drilling, an option which is not open to Western Europe or Japan.

Thus, one key effect of the changes in the international oil industry is a drastic shift in economic power from Western Europe and Japan to the United States. One might say that virtually all of the efforts of Western Europe and Japan over a twenty-five period to build up vast financial power—in the form of gold and dollars—at the expense of giving up much real physical investment to U.S. multinational corporations could be nearly wiped out at a stroke. That this is no mere hyperbole can be seen from the fact that the projected increase in Western Europe's import bill for oil in 1974 alone is greater than the total U.S. investment in Western Europe built up over these twenty-five years. Even Europe's gold reserves, which have an official value of $25 billion, but a free market value of about $100 billion, would not last very long in the face of these huge oil bills.

In response to this situation, each developed country has been frantically and unilaterally seeking its own way out. One method has been for each country to seek to increase its exports, but since most of the developed countries' trade is among themselves, this clearly cannot be a solution for all of them. Another route has been to negotiate direct government-to-government barter deals with individual OPEC countries, swapping a range of developed country goods for crude oil. For example, by early 1974 France had tentative agreements with Saudi Arabia, Abu Dhabi, and Iran to trade Mirage jets and petroleum refining and industrial equipment for crude oil; similar deals were also announced for Britain, West Germany, and Japan. However, for reasons discussed below, there are many barriers to a large number of such deals being finally consummated. Nevertheless, they do pose serious problems for the United States, since presumably in such direct government-to-government deals the oil companies would not get their usual profits. This would then cut into the protective cushion the American companies form for the U.S. balance of payments.

Assuming the OPEC countries, and Saudi Arabia in particular, do take in the kind of money flows indicated here, how can the West avoid the balance of payments problem inherent in this situation? As we have indicated, military equipment is one possible partial solution, but particularly for a country like Saudi Arabia with a population of only seven million, mostly uneducated, this would seem to be a limited market. Similarly, the market for Western industrial plant and equipment would also be limited, for lack of labor and other natural resources aside from energy; moreover, even if large-scale foreign investment took place in Saudi Arabia this would afford only temporary relief because production from these plants would largely have to be exported abroad and would further add to the balance of payments problems of Western Europe and the United States.

Another theoretical possibility is that the OPEC countries, and again Saudi Arabia in particular, would simply sit on their ever-growing foreign exchange reserves, investing them perhaps in gold or low-paying treasury bonds of foreign countries. It seems to me highly unlikely that any OPEC government, including Saudi Arabia, would agree to what in effect would be a "sterilization" of its assets, since it would derive no benefit from the situation other than increasing numbers of gold bars or pieces of paper. Moreover, there would always be a potential danger for the West: if the OPEC countries decided to use these reserves in a more vigorous fashion they could cause it severe damage.

The final prospect which has been talked about is that the OPEC countries, and once more Saudi Arabia in particular, could invest their vast foreign exchange reserves in buying up foreign stock and companies abroad. It seems to me that such a procedure would be totally unacceptable to the West. Thus, the $400-$500 billion which the OPEC countries theoretically could amass by 1980 could buy up close to half of all the corporations in the United States! Aside from being a fantastic transfer of U.S. wealth, which took over 200 years to build up, it would just increase the U.S. balance of payments problem even in the medium run. Thus, an investment of $400 billion which yields a 10 percent per annum profit would simply be another $40 billion per year drain on the U.S. economy. The result of all this would be that the United States would become like an underdeveloped country in respect to the OPEC nations. Such a situation undoubtedly would be unacceptable; it would be much cheaper to invade and take over the OPEC countries, Saudi Arabia in particular.

Therefore, we are left with the following conclusion. At present prices of oil, in the next few years Western Europe and Japan would be badly hurt while the United States would be greatly strengthened vis-à-vis these countries. However, since the United States has a big stake in investment and trade with Western Europe and Japan, a collapse there is not to U.S. advantage and hence to be avoided. It is quite conceivable, however, that the U.S. strategy is to try and walk the tightrope of benefiting from Western Europe's and Japan's weakened position in the short run while not allowing the latter's situation to deteriorate to complete collapse.

In this connection, it is interesting that the authoritative London *Economist*, in a July 1973 article entitled "The Phoney Oil Crisis," commented on the 1970–1971 negotiations with OPEC as follows:

> Yet the suspicion remains that the American State Department caved in too easily, though to ascribe this to simple blundering is not convincing either. It is more likely that there was another element, or elements, in the situation which have not been made public, and perhaps never will be. One suspicion is that some concern from a higher level of politics affected the way the Americans acted . . . the fact that the suspicion exists in the industry is significant, as is the guess, widespread in some places, as to at least one of the reasons for the Administration's attitude. According to it, the Americans gave in to OPEC so readily because they saw increased oil prices as a quick and easy way of slowing down the Japanese economy, whose exports were bothering America mightily at the time and which would be more hurt by rises in oil prices than any other nation.

Now, after the October War, the United States is in an even better position to sit back and watch the treasuries of Japan and Western Europe being relatively quickly emptied, figuring to share these riches with the OPEC countries. If and when a point is reached where oil prices threaten to harm the United States, either directly or through a collapse of the other developed countries, then it could use its power to try and pressure the OPEC governments to reduce prices, with the ultimate threat being military intervention.

However, since military intervention would raise the danger of confrontation with the Soviet Union as well as of prolonged guerrilla warfare in the Middle East, it will not be chosen lightly. But it is a very real possibility. Trial balloons were sent up after the October War by everyone from Senators Fulbright and Jackson to Secretary of State Kissinger to Secretary of Defense Schlesinger. But with the American people sick of war and suspicious of the oil companies and the president (according to the polls at that time the majority of people blamed the oil crisis on the oil companies and the government and not the Arabs), intervention may not have been tenable. Instead, such warnings may have been directed at preserving the privileged position of U.S. oil companies in the Middle East. That is to say, the saber-rattling may have been a not so subtle way of warning the OPEC governments not to truly nationalize American oil companies in the Middle East nor to set prices so as to reduce the companies' profitability sharply.

In any event, this is a most tricky and dangerous game for the United States, threatening to plunge the world into a war or a major depression. Ironically, a major economic crash in the Western world would also help "solve" the oil problem by drastically reducing the demand for oil and hence putting strong downward pressure on the price. Since the developed countries have long been teetering on the edge of a financial breakdown which would precipitate a major world depression, the international oil crisis might yet be just the factor necessary to tip the world economy over the precipice. After all, the classic capitalist solution to both inflation and mal-distribution of income has been depression, and nothing has changed on the world scene to indicate the degree of coordi-nation necessary among capitalist powers to prevent the re-currence of such a depression.

This is not to say that the OPEC countries can or should try to solve the problem by reducing their oil prices. For one thing, the "oil crisis" is just the latest stage in the continuing deterioration of the underlying economic structure of the world capitalist system. As leading economist Peter Bernstein has noted, we have reached a point where:

> We still face the risk of a 1929-type worldwide financial smash...
>
> Most of the necessary conditions for it are present: a banking system with a thin capital structure and many illiquid assets, an explosion in debt by most businesses (including many of the largest and most prestigious), a complex Eurodollar credit system without any lender of last resort, a simmering financial crisis in the real estate area, a multinational scramble to accumulate inven-tories at swollen prices and the balance-of-payments crisis for many nations in the wake of the oil embargo of last winter.
>
> Inflation itself creates continuing financial strains, while a trend in the other direction would leave many firms unable to meet obligations incurred on the assumption that they would be able to raise or at least hold their selling prices.

The OPEC countries do not have the power to stop this process of decline, because its roots, as we have seen, run much deeper and lie in the basic rivalries and lack of coordi-nation in the developed capitalist world. Given the likelihood that an economic deluge is in the cards regardless of what OPEC does, the only rational course for its members is to pile up as much money and resources as possible aboard their own Noah's Ark.

Moreover, in a fundamental sense what is involved as a result of increased OPEC crude oil prices is the long overdue need for a sizable shift of real resources from the developed countries to the oil-producing countries. The huge economies of the developed countries are certainly capable of making these resources available. Thus, while the balance of payments problems are undoubtedly very serious under existing institutional arrangements, new institutional arrangements could be found to allow this shift of real resources if the developed countries had the will to do so.

After all, if Saudi Arabia receives $25 billion per year, this amounts to a per capita income of about $3,000 per person, which is well below that of the United States. Considering that the Saudi infrastructure is so undeveloped, in fact it would take a number of years before even a much higher income level would bring the Saudis anywhere near the U.S. standard of living. But, more fundamentally, within the logic of capitalist economics, why shouldn't the Saudis have a U.S. standard of living, or a higher one, since they have physical resources in the form of crude oil which have great value to the world?

Moreover, if one takes a broader view and counts the considerably more than 100 million people in all the Arab countries or the several hundred million in OPEC member countries as a whole, then even $200 billion per year for OPEC would amount to less than $1,000 per person. The OPEC countries could easily absorb millions of prefabricated houses and other durable consumer goods, capital equipment, and agricultural products, along with personnel such as teachers and engineers, to assist in the development of their indigenous economic capacity.

Even if the OPEC countries were unwilling to accept huge flows of goods and personnel in the early years because of fear of disrupting their economic, political, and social systems, the developed countries could find ways to guarantee that someday when the OPEC countries are ready, enormous flows of real goods and resources would be made available. In other words, as in the United States, when it is felt that it is a matter of life or death to develop something like the atomic bomb, tremendous resources can be concentrated upon dealing with the problem. Similarly, if it were truly accepted by the developed countries that the only way to obtain the crude oil which is vital for their economies and standard of living is to provide a large flow of real resources to the OPEC countries, then undoubtedly the developed countries can find ways to make such arrangements.

Thus, in the final analysis, OPEC should not be deterred from seeking fair prices, based on the value of its oil to consumers, by the inadequacies of present institutional arrangements. Otherwise these inadequacies will be used as an excuse for not making possible the real transfer of resources which both equity and the new world power situation demand.

However, these considerations of "morality" touch on questions which are the concern of the next chapter. The hard lesson of this chapter is that once again the world imperialist system, the most advanced form of capitalism, ultimately offers the world its two traditional and horrendous solutions to its fundamental crises—war or depression. The analysis of how energy might be used in a rational world, which we turn to next, happily provides some bright relief from this dismal picture.

9
Energy in a Rational World

In this book we have examined the historic and current roles of the principal forces involved in the international energy crisis, as well as their likely future roles barring any major institutional changes. In this final chapter we turn to an examination of how the international energy crisis in theory could be overcome, and more generally, how energy resources could be used ideally in a rational world. Such an examination highlights the absurdity of the present world capitalist system of energy resource use which most of us have come to take for granted as the "normal" or "efficient" system.

Any attempt to specify a rational method of energy resource use requires a prior specification of the principles which would underlie a rational system for using all economic resources, of which energy would be only one important component. Here the gap between the actual and the possible becomes dramatically apparent. But only when the links between the overall economy and the energy sector are clear can we begin to see the extensive social changes which must take place in order to overcome the energy crisis and develop a rational basis for energy resource use.

Thus, in this chapter we will compare two situations. The first is an ideal model of the kind of international economy and energy resource use which would take place in an integrated world community devoted to meeting the needs of all of its people. The second is the actuality of the international capitalist economy and its energy resource use. From this comparative analysis I believe some useful lessons can be drawn about what needs to be done to solve the world's energy problems.

To begin with, in my view the fundamental goal of a rational world economy would be to move toward increasing equality among all people. Not only is such a goal desirable in terms of equity and justice, but it is also necessary in the long run if the world is not to tear itself to pieces.

From this overall principle several other general principles of economic resource allocation follow. First, economic resources should be used to provide for an equal distribution of income among countries, sexes, races, etc. Second, and perhaps less obvious, economic resources need to be used to provide relative equality between future and present generations. More specifically, individual income should increase over time only by that which would be allowed by improvements in technology and productivity, less resource depletion. Furthermore, the environment must be considered as a resource for which the present generation also has obligations to future generations. Finally, a third principle, which will help make it easier to act upon the first two, is that economic resources should be used for production of necessities and comforts, but not wasteful luxuries. (The line between comforts and luxuries is partly defined by the overall availability of economic resources.)

As for the principles underlying rational use of energy resources, this is part of the general problem of rational use of nonreproducible resources, particularly minerals. One principle is that there would be an international division of labor so that the lowest cost resources would be utilized first. (Until a truly integrated world community is achieved, of course, this principle must be modified by the need of individual countries to have some degree of self-sufficiency in production so as to insure their independence.) That is, the world's oil should be provided by the Middle East, the world's coal by the United States, etc. In making these decisions, costs would be measured basically by the amount of labor needed to produce the resources, both directly and indirectly through the labor needed to provide the capital equipment used in energy production as well as transportation.

Such an international division of labor would insure the most efficient use of energy resources for the present generation. However, it would not insure a fair division of energy resources over time, so that it must be constrained by another principle. More specifically, while it might be most efficient for today's generation to drain the Middle Eastern oil fields dry over the next fifty years, it would not be fair to future generations unless there was some clearly foreseeable replacement for this low-cost oil.

How do we solve this problem of equity among generations? One answer, which is implicit in the general principles of rational economic resource use, would be to cut out luxuries and waste. Thus, if we don't have gasoline-gobbling Cadillacs, and in general favor collective consumption over private consumption, such as in the transport area, fewer energy resources would be required. Even with today's technology, which has been aimed primarily at promoting automobile transportation at the expense of mass transport, it requires far less energy to transport an individual a given distance by bus or railroad than by automobile. And if the ancient railroad system with its heavy steel cars were modernized, energy consumption could be reduced even further.

More generally, however, it would take long-term planning (fifty to a hundred years), constantly being updated to take into account changes in resource availability and technology, to avoid inequities among generations. For caution's sake, there should be a limitation on the use of currently nonreproducible scarce resources. As we have said, no single generation or pair of generations has the right to use up all of a given resource unless there are replacements. If the present generation feels confident that improved technology will one day make this replacement possible, then it could legitimately increase its current use of the resource—providing, however, that if it does not discover this new technology within a reasonable time, it will repay future generations for its overoptimism. This could be done by then reducing its consumption and/or by spending increased amounts for exploration and technological research.

The basic idea is that at each time in its history society has to the best of its knowledge a given quantity of resources available. The possibilities of increasing the supply of resources depend on exploration to discover their location and technological research into developing substitutes. Equity among the generations requires that a baby born in 1975, with a life expectancy to 2050, should be able to assume the following. First, that it will get its share (one three-billionth) of current world income based on 1975 technology and resource use, and also one three-billionth of world resource reserves. Second, that by the year 2050 its income will have risen, as long as a newborn baby's share of known current world resources is no less than a baby's was in 1975.

What this means concretely, for energy resources, is that the world should first use up Middle Eastern crude oil, which costs ten cents per barrel. But if it appears that these supplies will be exhausted by 2050 and no replacement for oil is yet in sight, then the generation born in the second half of the twenty-first century—which will have to turn to higher-cost oil, say from the United States—should be compensated by the earlier generation. This can be done by requiring the earlier generation to cut back on the use of coal, or to reduce its consumption of products requiring large amounts of energy, such as aluminum, or by spending more resources to find oil substitutes.

In striking contrast to these principles of rational resource allocation we have the present irrationality of the world capitalist system. Here use of economic resources is based on the existing distribution of wealth and income, which within and among countries is extremely unequal. Moreover, little consideration is given to the interests of future generations except very indirectly through interest rates determined in financial markets, as will be discussed below. Within a country an individual's share of economic resources is heavily based on the accident of birth, which determines his or her sex, race, health, and family wealth and income. Among countries the distribution is largely based on accidents of geology or history, also to be discussed more fully below.

The overall resource allocation system is irrational fundamentally because it has nothing to do with need—in fact, the opposite is more often true, since the more needy one is, the less one usually gets. This is partly mitigated in Western developed countries by social welfare reforms which tend at least to keep relative income shares stable, even if the absolute gap between the rich and the poor grows wider and wider. However, the irrationality is not mitigated in the relations between the developed and the underdeveloped countries, for there is a growing income gap not only absolutely but also relatively. Thus, for example, while per capita income in India in 1950 was one-thirtieth that of the United States, by today it is about one-fiftieth.

This irrationality also extends to the treatment of future generations. Capitalist market theory says that this problem is solved by the free market interest rate. When this rate is high people are assumed to defer present consumption and to save, so that more resources will be available for future generations. Unfortunately, however, in the real world, this interest rate has little real effect. For one thing, it doesn't help in the many cases where actions of business have a cost to society but none to the firm. The classic case, of course, is pollution, where the cheapest way for a business to get rid of waste products may be to dump them in the nearest river, thereby fouling it for future generations. While this is now widely noted, it is also true that unemployment, war, racism, and sexism are phenomena which corporations frequently contribute to at great cost to society in terms of human and material resources, but for which they pay no penalty themselves.

Furthermore, the interest rate theory does not take into account that resource decisions are made initially and primarily by private companies, not by consumers; the latter are then high-pressured by advertising and other methods into ratifying the companies' production decisions. Moreover, since private companies normally seek a rate of profit far greater than the going interest rate, this target rate of profit is the real mechanism for allocating resources between present and future generations; and because the target rate is quite high, companies normally put a premium on quick plunder now, leaving relatively little for the future. (This is even more true of foreign corporations in underdeveloped countries which seek even higher profits and quicker rates of payback out of fear of nationalization.) Thus, as opposed to the ideal that government should give equal consideration to all of its generations over the years, international capitalism poses the reality that its emphasis on private consumption as the be-all and end-all of life creates tremendous drives for immediate consumption (which is also constantly needed to prevent the system from collapsing into depression), and disregard for the future.

The economies of the underdeveloped countries are a magnification of the worst characteristics of capitalist resource allocation, without any of the mitigation made possible by the wealth of the developed capitalist countries. The underdeveloped countries are underdeveloped not by accident but because of the activity and influence of the developed countries. Largely because of peculiarities of geography and history, development began in Western Europe, particularly in Great Britain. Once this process began, Western Europe then systematically plundered the resources of the rest of the world and blocked development that threatened competition with it. For example, Great Britain forced opium into the bloodstream of China in order to gain China's tea and silver, and in India it destroyed much of the indigenous textile industry in order to provide markets for British textiles. In the United States, which again largely through accident was enabled to break free of British power, much of the economy was built on the slave labor of Africa.

The economic conquest of the "underdeveloped world" began so long ago that few people realize that many presently underdeveloped countries were relatively well developed before the onslaught of the West. For example, Chile had a flourishing copper industry in the late nineteenth century which was entirely owned by its own citizens. Today its ownership by developed countries, and the United States in particular, is maintained through classic imperialist manipulation and overthrow of governments.

But even within the poverty left after the developed countries have looted the underdeveloped ones, still there is tremendous inequality of income. This inequality is often fostered by the corporations of the developed countries in order to keep native labor toiling under inhuman conditions. For example, the United Fruit Company and the aluminum companies in the Caribbean were notorious for holding more land than they could use so that there would be landless workers who would have to toil on their plantations and in their mines (this is in principle no different from the "head tax" or "hut tax" imposed on inhabitants of Southern Africa which drives them to work in the mines). This enormous inequality also serves to buy up and help keep in power the corrupt indigenous leaderships of many underdeveloped countries; the landlords and commercial classes need a rich living standard in order to be satisfied with their place in the imperialist sun, while the military needs that plus all kinds of weapons and equipment.

In the final analysis, then, the underdeveloped countries are merely the logical culmination of capitalist resource allocation. They are the prime examples of capitalism's proclivity for what has been described as "building on the best," which in effect means that the rich get richer and the poor get poorer. While the underdeveloped countries are at the bottom of the heap, still they have within their cities tiny enclaves of luxury which get more and more developed while the rest of the underdeveloped world becomes an increasingly vast wasteland of human misery.

The capitalist allocation of energy resources in particular is a classic example of this irrationality. The developed countries, on the one hand, get most of the energy resources because they have most of the money (and in times of scarcity, such as after the October War, they can bid away these resources from other underdeveloped countries). For years the developed countries used these energy resources incredibly wastefully because oil was a relatively cheap good. Thus, because cheap gasoline in the United States made big cars economically feasible for many, private transportation was the road chosen instead of fuel-efficient mass transport. (Gasoline taxes paid for a huge highway system which was built after World War II.) This era of waste in turn has led to enormous environmental problems like pollution and the paving over of America.

With the "cheap oil" which could be increasingly extracted from abroad, there was no incentive for research for non-polluting energy sources, such as solar or wind energy. Moreover, since it obviously would be difficult if not impossible for the international oil companies to monopolize the sun or the winds, such research would clearly be to their disadvantage. In fact, even much more mundane research on turning coal into liquefied gasoline, which appeared technologically possible many years ago, was blocked by the power of the international oil companies.

Only now is some thought being given to conservation of energy, through such methods as recycling the enormous heat given off by the air-conditioners, used to cool banks of computers, to heat the rest of the building. Even now there is tremendous resistance in the private sector to adapting such simple techniques as requiring collection of cattle manure, which either can be used as fertilizer, thereby saving petroleum needed for petrochemicals, or can be dried and used as a fuel directly; over a billion tons of animal manure are produced in the United States annually, and most of this is allowed to be washed into rivers, polluting them as well as losing the potential benefits of the manure. All this simply because it would be costly for cattle owners to collect the manure.

As for capitalist energy resource allocation in the Third World, most of the underdeveloped countries have likewise become heavily dependent on imported crude oil. This has happened partly because imported crude was cheap, and also because the unholy combine of the international oil companies, their home governments, and the international financial institutions have worked assiduously to block them from seeking their own oil. Now, after the October War, many of these underdeveloped countries are in the disastrous position of being unable to afford even the quantities of oil which they previously imported, or the petroleum-based fertilizers they so badly need—fertilizers which are not produced locally partly because they lack their own oil industry. As a result, countries like India are now going downhill economically at a fantastic human cost.

On the other hand, a handful of oil-exporting underdeveloped countries for many years got a relative pittance while their precious and in many cases sole natural resource was wastefully sucked from the ground. Now, after the October War, a handful of them will get relatively vast riches. Yet, within the framework of capitalist economic relations they will not find a useful outlet for these riches. For one thing, without a social revolution at home they can't use these funds productively for general social benefit. Thus some amount will be spent on foreign investments, partly because dictators, like squirrels, are notorious for storing wealth in safe places for a rainy day (witness the enormous sums looted by Batista of Cuba and Gómez of Venezuela; more recently, Shah Pahlevi's "foundation" in New York has invested millions of dollars of Iran's oil wealth in New York real estate).

But, as we have seen previously, foreign investment for the oil-producing countries has its limits. And, of course, the few oil-producing countries which have far more revenues than their people can effectively utilize (e.g., Saudi Arabia, Kuwait) cannot donate these funds to other underdeveloped countries which might use them more effectively, because it would violate the capitalist ethics (two eyes for an eye, a mouth for a tooth) of the dominant class in these countries, many of whose members have been trained at Western business schools.

Moreover, even if the oil-producing countries were more benevolent, in a world of capitalist irrationality such a gesture would be futile. It would be like an individual trying to solve the problem of poverty in America by giving all his money to beggars. In order for other underdeveloped countries to truly begin the task of economic development, it will not suffice to pour large amounts of money into societies which will ultimately waste them because of their own misshapen economies and social structures.

Our general conclusion, then, is as follows. Within the capitalist world, not only is there vast inequality of wealth and income, but basic mechanisms exist to insure that the gap between needs and resources, between poverty and opulence, both within and among countries, will never be bridged. Only a genuine social revolution within each country can make possible the rational use of its own economic and energy resources. And only a genuine social revolution in every country can make possible the rational use of all resources on a worldwide basis.

Each country which undergoes such a social revolution and opts out of the world capitalist system contributes to this process in two ways. Not only does it bring another group of people into the world of rational resource planning, but by reducing the scope of the capitalist world it increases the tensions and contradictions within it, thereby speeding the process by which the whole system becomes increasingly dysfunctional and unbearable for those who live within it.

This is not to say, however, that constructive measures cannot be taken now by people in the capitalist world, both developed and underdeveloped, to move toward a more rational allocation of economic resources in general and energy resources in particular. As far as the underdeveloped countries are concerned, the vast transfer of wealth from the developed countries to the OPEC countries following the October War is itself a tremendous victory which moves in the right direction of decreasing inequality among countries. As this wealth accrues in the OPEC countries, to some extent it will help improve the standard of living of the people and for this we can all be thankful. More important, however, since this wealth is unlikely to be effectively utilized, its existence will call forth demands for change and improvement within the OPEC countries, which in my view will ultimately lead to a social revolution; as education and communication spread, and the gap between rich and poor and the possible and the actual grows greater, the pressures for radical change will be irresistible.

Furthermore, the OPEC victory may be a stimulus to other underdeveloped countries which have important irreplaceable resources such as tin, copper, and bauxite to join forces in trying to get more equitable and higher prices for these commodities. And, if they are to do so, these countries will have to challenge the power of the multinational corporations which now control their resources. Such a challenge if successful will weaken both the imperialist forces in their home countries, thus increasing tensions in the capitalist world, and the imperialist hold on the underdeveloped countries, thereby allowing room for further radical change within these countries. (Even if the challenge fails, it would be extremely instructive to the people of the underdeveloped countries to become aware that they are not allowed to control their own resources.) Only in the course of struggling to break the bonds of imperialism can the people of an underdeveloped country get a real education about the internal forces which are in league with imperialism in blocking the country's development.

As for the developed countries, many proposals have been made for governmental action to "solve" the energy crisis. Within the United States these have ranged from such patent cosmetics as Senator Jackson's "federal chartering" of energy companies all the way to complete nationalization of these companies. While the latter could be a useful step, it should be clear that it would not solve the energy problem.

It could be useful because theoretically it would mean public control over a vast power center in our society, one which has historically led to enormous foreign intervention and domestic resource waste. Moreover, the struggle to make

this change, by bringing into opposition the enormous powerful forces of capitalism, would itself be extremely educational as to the real nature of wealth and power in our
society.

On the other hand, as long as the rest of the economy and
society are run on the basis of capitalist market principles,
which misuse resources in general, even a nationalized energy
sector would not prevent a continual misallocation of energy
resources. That is, as long as private autos are the dominant
form of transportation, even a public energy corporation
would have to provide wasteful amounts of gasoline. And if
the energy resources are not available here, the public energy
corporation would have to seek them out abroad just as the
private companies do now. Moreover, even if they were available here, if the U.S. balance of payments situation required
profits from U.S. ownership of foreign oil, the public energy
company might be even more aggressive abroad than the private ones. In short, a public energy company would at most
be a more rational mechanism for meeting the demands of a
fundamentally irrational economy. As such it could turn out
to be either more or less harmful for the social good, depending on circumstances.

In a real sense, then, the problem of the "energy crisis" is
itself simply a reflection of far deeper underlying irrationalities of the world capitalist system. The true core of the problem is that the motive force of capitalism is not human needs
but the drive for profits. As Adam Smith so aptly put it 200
years ago, "It is not from the benevolence of the butcher, the
brewer, or the baker, that we expect our dinner, but from
their regard to their own interest." While this view has long
been trumpeted by supporters of the system as a triumph of
rational thinking over fuzzy-minded idealism, in fact its profound negative implications are now coming to be more
widely understood.

When combined with the unequal distribution of income
within the society, the centrality of the profit drive means
that for both short-run stability and long-run growth, profits
and the capital investment which generates new profits must

be ever increasing. In order to make profits, however, goods and services must be produced and sold, and this in the long run means ever increasing consumption of raw materials, of which energy is one central component. Moreover, as capital equipment increasingly replaces human labor, even more energy is required.

The resulting drive for increasing production and consumption takes place irrespective of human needs. Thus the heart of the energy crisis is *not* that there are shortages of energy resources in the world or in some countries, or that the price is "too high." The real heart of the crisis is that in the world capitalist system there is no way in which human needs and desires, rationally balanced off against limitations of human and material resources, can act as the basic determinant of production and use of all goods and services, including energy. Instead of a social determination of needs and priorities now and in the future, the system plunges headlong along the roller coaster of production-consumption possibilities laid out by the mindless forces of supply and demand. It is therefore little wonder that many see life on Spaceship Earth as ticketed for misery if not extinction on a despoiled and polluted shell.

In the final analysis, in the developed as well as the underdeveloped countries, one reaches the same conclusion: there is no solution to the energy crisis short of a fundamental social revolution which dethrones the role of wealth and money and organizes the economy to meet the real needs of all the people. Only a thorough housecleaning of the total economy, not least in the vital energy sector, can do the job and insure that we meet our obligations not only to ourselves but to future generations. It is time for all of us to roll up our sleeves and begin this long overdue task, with all the energy we can muster. History will honor those with the vision and courage to undertake this desperately needed effort.

Notes

Chapter 1

13-14   The chart is reprinted from U.S., Congress, Senate, Committee on Interior and Insular Affairs, *A National Fuels and Energy Policy Study: Hearings on S. Res. 45* (hereafter referred to as *Jackson Committee Hearings*), 92nd Cong., 2d sess.; published as "Natural Gas Policy Issues: Part 2," Serial No. 92-22, February 25, 29, and March 2, 1972 (Washington, D.C.: U.S. Government Printing Office, 1972), pp. 969-70.

13-14   For a general discussion of the "ages of oil," see Harvey O'Connor, *The Empire of Oil* (New York: Monthly Review Press, 1955), pp. 8-18.

14-16   For data on shares of energy consumption, see: for 1929 and 1939, American Petroleum Institute, *Petroleum Facts and Figures*, 1971 edition; for other years, United Nations, Statistical Office, Statistical Papers, *World Energy Supplies* (series J).

16   For estimates of Middle Eastern and world proven oil reserves, see: for 1940, Zuhayr Mikdashi, *A Financial Analysis of Middle Eastern Oil Concessions, 1901-1965* (New York: Praeger, 1966), p. 91; for 1973, U.S., Congress, Senate, Subcommittee on Multinational Corporations, Committee on Foreign Relations (hereafter referred to as *Church Committee Hearings*), 93rd Cong., 2d sess.; published as "Multinational Corporations and United States Foreign Policy," Part 4, January 30, 1974 (Washington, D.C.: U.S. Government Printing Office, 1974), pp. 197-98. Note that reserve estimates for Saudi Arabia range from about 150 billion barrels to 300 billion and up.

16-17   For data on the average cost of production, see M.A. Adelman, *The World Petroleum Market* (Baltimore: Johns Hopkins University Press, 1972), p. 76.

17   For the Big Seven's share of the non-Communist world's oil reserves and refining facilities, see U.S., Congress, Senate, Select Committee on Small Business, staff report to the Federal Trade Commission, *The International Petroleum Cartel* (Washington, D.C.: U.S. Government Printing Office, 1952), pp. 5-6, 23, 25.

18   Walter Levy is quoted in Joyce and Gabriel Kolko, *The Limits of Power* (New York: Harper & Row, 1972), p. 447.

18-19   Taki Rifai, *The Pricing of Crude Oil* (New York: Praeger, 1974), pp. 207-8.

20   The director of the government task force is quoted in the *New York Times*, February 10, 1974, p. 42.

20   William Nordhaus, "The Allocation of Energy Resources," *Brookings Papers on Economic Activity*, no. 3, 1973.

21   For data on coal reserves and production, see: for all but Great Britain, U.S. Department of the Interior, Bureau of Mines, *Minerals Yearbook 1971*, vol. 3: *Area Report International* (Washington, D.C.: U.S. Government Printing Office, 1973); for Great Britain, United Nations, *Statistical Yearbook 1971*.

21   Estimates of the cost of oil imports in 1974 are based on the *Wall Street Journal*, June 6, 1974, where OPEC exports for 1974 are projected at $105 billion.

Chapter 2

23   United Nations Interregional Seminar on Petroleum Refining in Developing Countries, New Delhi, January 22-February 3, 1973.

23   O'Connor, *The Empire of Oil*, p. 10.

24-27   This history of the early years of the international oil companies draws heavily on ibid., pp. 10-18, and on Harvey O'Connor, *World Crisis in Oil* (New York: Monthly Review Press, 1962), chapters 1-7.

27   For the Big Seven's share (individually and totally) of the non-Communist world's crude oil production and refining capacity in 1949, see *The International Petroleum Cartel*, pp. 24-25.

27-28   Data for 1972 and 1973 are from company annual reports and from *Fortune* magazine; the 1947 and 1949 rankings of company profits and assets are from U.S., Congress, Senate, Committee on the Judiciary, Subcommittee on Antitrust and Monopoly, *Economic Concentration: Part 1: Overall and Conglomerate Aspects: Hearings on S. Res. 262*, 88th Cong., 2d sess., July 1, 2, and September 9, 10, 11, 1964.

28   For world oil reserve data, see: for 1949, Mikdashi, *A Financial Analysis*, p. 91; for 1973, *Church Committee Hearings*, pp. 197-98. For Middle East and world production data for 1973, see *The Petroleum Economist*, March 1974, p. 86.

29   On the growth of vertical integration in the United States oil industry, see John G. McLean and Robert William Haigh, *The Growth of Integrated Oil Companies* (Boston: Harvard University Press, 1954).

30   For data on the oil companies' foreign production, see: for 1939, *Moody's Analyses of Investments: Industrials* (1942); for 1972, First National City Bank, *Energy Memo*, April 1974.

31   For the oil companies' share in petrochemicals in 1962, see Jules Backman, *The Economics of the Chemical Industry* (Washington, D.C.: Manufacturing Chemists Association, February 1970), p. 73.

33   The Exxon official is quoted in James Ridgeway, *The Last Play: The Struggle to Monopolize the World's Energy Resources* (New York: Dutton, 1973), p. 54.

33   Senator Aiken's testimony is from U.S., Congress, Senate, Committee on the Judiciary, Subcommittee on Antitrust and Monopoly, *Competitive Aspects of the Energy Industry*, 91st Cong., 2d sess., May 5-7, 1970, pp. 8-9.

34   The Exxon prediction on coal production for synthetic gas is cited in Ridgeway, *The Last Play*, pp. 75-76.

34   Uranium data are cited in *Competitive Aspects*, pp. 118-20.

34-35   On the origins of the "energy crisis," see Robert Sherrill, "Energy Crisis: The Industry's Fright Campaign," *The Nation*, June 26, 1972, pp. 816-20. The Supreme Court opinion is quoted on p. 817.

35   Ridgeway, *The Last Play*, p. 134.

Chapter 3

39   Exxon's treasurer is quoted in Robert Engler, *The Politics of Oil* (New York: MacMillan, 1961), p. 267.

39-40   Ibid., pp. 310-12.

40-41   *Wall Street Journal*, June 17, 1974, p. 28.

42-46   This discussion of the history of the international oil companies draws heavily on O'Connor, *World Crisis in Oil*, chapters 3, 23, 25-27, with specific quotations in the following order: British officer, p. 279; British government view, p. 282; U.S. Secretary of State cable, p. 284; U.S. Ambassador to Iran, p. 287; Exxon president, p. 305; State Department and British Foreign Office statements, p. 306.

46   The U.S. State Department memo is quoted in Engler, *The Politics of Oil*, pp. 191-93.

47   On U.S. penetration of former European colonies, see: Victor Perlo, *American Imperialism* (New York: International Publishers, 1951), pp. 37, 172-91; and Kolko, *The Limits of Power*, p. 448.

47-48   The U.S. Ambassador to Bolivia is quoted in Kolko, *The Limits of Power*, pp. 416-17.

48   For data on the U.S. share of Middle Eastern reserves in 1940-1950, see Mikdashi, *A Financial Analysis*, p. 91.

48-51   For a detailed discussion of the Iranian oil events, see Michael Tanzer, *The Political Economy of International Oil and the Underdeveloped Countries* (Boston: Beacon Press, 1969), pp. 321-26.

50   The quote is from Hossein Sheikh-Hosseini Noori, *A Study of the Nationalization of the Oil Industry in Iran* (Ann Arbor: University Microfilms, 1965), pp. 248-53.

50-51   David Wise and Thomas B. Ross, *The Invisible Government* (New York: Random House, 1964), p. 110.

51   Anthony Eden, *Full Circle* (Boston: Houghton Mifflin, 1960), p. 647.

51-52   The *New York Herald Tribune* and State Department reports are in David Horowitz, *The Free World Colossus* (New York: Hill and Wang, 1965), p. 191.

52   The *New York Times* report and the Engler comment are in Engler, *The Politics of Oil*. The Dulles statement is in C. Wright Mills, *The Causes of World War III* (New York: Simon and Schuster, 1958), p. 66 (emphasis added).

52   On the Kassem overthrow, see Horowitz, *The Free World Colossus*, p. 192.

Chapter 4

54-58   George W. Stocking, *Middle East Oil* (Kingsport, Tenn.: Vanderbilt University Press, 1970).

55   Ibid., pp. 3, 11, 31, 52.

55   O'Connor, *World Crisis in Oil*, p. 128.

56   Stocking, *Middle East Oil*, pp. 5-6, 16.

56   *Special Report by His Majesty's Government to the Council of the League of Nations on Progress of Iraq, 1920-1931* (London: British Colonial Office, 1931), p. 14; quoted in ibid., p. 51.

56   Ibid.

56-57   Edwin Lieuwen is quoted in O'Connor, *World Crisis in Oil*, p. 135.

57   Data on the Persian concession is from Mikdashi, *A Financial Analysis*, pp. 9-21, 73-78. Since prices of Persian Gulf crude oil were not published before 1950, I have estimated its value by assuming it to be roughly equal to the average price of crude oil at the well head in the United States.

57   Data on the Iraqi concession are from ibid., pp. 65-73, 83; data on the Venezuelan concession is from O'Connor, *World Crisis in Oil*, p. 129.

57   Stocking, *Middle East Oil*, p. 130.

58   Ibid., pp. 130-31.

58   Data for the Middle East from 1900-1960 are from Charles Issawi and Mohammed Yeganeh, *The Economics of Middle Eastern Oil* (New York: Praeger, 1962), p. 108.

58-59   Data on historic profitability in Iran and Iraq are from Mikdashi, *A Financial Analysis*, pp. 221, 194, 272-82; my calculations of future profits and peak foreign seed capital invested in Iraq are based on Mikdashi's Table 34, pp. 275-76.

59   For a fuller treatment of direct Western government intervention to assist the oil companies, see Michael Tanzer, *The Sick Society: An Economic Examination* (New York: Holt, Rinehart and Winston, 1971), pp. 82-89.

59-60   On U.S. intervention in Guatemala, see ibid., pp. 84-86.

60   On the role of international organizations in the oil area, see Tanzer, *The Political Economy of International Oil*, pp. 90-106.

61   *Wall Street Journal*, June 24, 1959, p. 1.

61-62   For detailed analyses of the origins and metamorphoses of these resource flows, see Andre Gunder Frank, *Capitalism and Underdevelopment in Latin America* (New York: Monthly Review Press, 1967).

63   Sheik Abdullah Tariki is quoted in David Hirst, *Oil and Public Opinion in the Middle East* (New York: Praeger, 1966), pp. 57-58.

65-66   Ibid., pp. 100-102.

66   On OPEC's 1960 goals, see Hussein Abdallah Abdel-Barr, "The Market Structure of International Oil with Special Reference to the Organization of Petroleum Exporting Countries" (Ph.D. diss., University of Wisconsin, 1966); quotation from p. 56. The defection of Royal Dutch Shell from the other majors may be attributable to the fact that it was the weakest of them in terms of its ownership of Middle Eastern crude oil.

67   On specific price information, see ibid., p. 49.

67   On OPEC's relative lack of success, see Hirst, *Oil and Public Opinion*, pp. 114-15.

68   On divisions among OPEC members, see ibid., pp. 112-13. Thus it was reported that at the Sixth Arab Oil Congress, "The ideological split between the Socialist-style and monarchist regimes—or between 'nationalists' and 'moderates'—prevented the conferees from advancing any closer to a unified or coordinated oil policy . . ."—*Petroleum Intelligence Weekly*, March 20, 1967. Hirst also notes (ibid., pp. 103-104) that: "On the whole, it is a moderate and realistic public opinion to which the technocrats have always sought to make their appeal. For to advocate extreme measures would bring them into conflict with the political executive in their home countries."

68   *Petroleum Intelligence Weekly*, March 20, 1967.

69   On OPEC's appeal to world public opinion, see Hirst, *Oil and Public Opinion*, p. 114.

69   OPEC, "OPEC and the Principle of Negotiation," paper presented to the Fifth Arab Petroleum Congress, Cairo, March 1965, pp. 18, 19.

70   Estimate of total OPEC revenues in 1974 is from the *Wall Street Journal*, June 6, 1974, p. 1; individual country estimates are from the World Bank, as reported in *The Petroleum Economist*, May 1974, p. 165.

72   *New York Times*, January 16, 1974.

72-77   The analysis of the situation in Yemen draws heavily on:

Joe Stork, "Socialist Revolution in Arabia: A Report from
the Peoples Democratic Republic of Yemen," *MERIP
Reports*, no. 15, March 1973. See also Eric Rouleau,
"Revolutionary Southern Yemen," *Monthly Review*, May
1973, pp. 25-42.

73   Stork, "Socialist Revolution," pp. 3, 5.
75   Ibid., p. 8.
76   Ibid., p. 23.
77   Ibid., p. 21.

Chapter 5

78-79   The discussion of Germany and oil before World War II
draws heavily on O'Connor, *World Crisis in Oil*, pp. 49-52,
67-68, 303-5; quotation from p. 49.

80-81   For a comprehensive analysis of the French role in oil, see
Farid W. Saad, "France and Oil: A Contemporary Eco-
nomic Study" (Ph.D. diss., Massachusetts Institute of
Technology, 1969); for data on French oil interests in
Algeria, see *The Petroleum Economist*, March 1974, pp.
88-89.

81-83   For the Iraqi case, see the various countries' diplomatic
"representations" as reported in the *New York Times*, May
16, 1967; State Department actions as reported in the *Wall
Street Journal*, February 15, 1974, p. 23; *Petroleum Intelligence Weekly*, October 23, 1967; *The Oil and Gas Journal*,
November 6, 1967, p. 69; *New York Times*, December 13,
1967.

84   Data on French-controlled crude oil are from *The Petroleum Economist*, March 1974, pp. 88-89.

84-85   For a discussion of ENI and Italian oil until 1946, see P.H.
Frankel, *Mattei: Oil and Power Politics* (New York:
Praeger, 1966), pp. 31-62.

85   Mattei is quoted in *World Petroleum*, February 1961, p. 40.

86   The comment on ENI and the underdeveloped countries
was made by Frankel, *Mattei*, pp. 128-29.

86-87   *Petroleum Intelligence Weekly*, October 25, 1965.

87-88   For information on Japan and oil, see: Richard O'Connor,
The Oil Barons (Boston: Little Brown, 1971), pp. 328-30;
Robert Guillain, *The Japanese Challenge* (New York:

Lippincott, 1970), p. 169; and J.E. Hartshorn, *Politics and World Oil Economics* (New York: Praeger, 1962), p. 287.

89   Data on U.S. oil investment in the Middle East at the end of 1972 is from U.S. Department of Commerce, *Survey of Current Business*, September 1973, p. 26.

90-91   According to an unidentified expert who commented on Lyndon Johnson's resignation following the March 1968 gold crisis: "The European financiers are forcing peace on us. For the first time in American history our European creditors have forced the resignation of an American president."—*Wall Street Journal*, April 4, 1968, p. 3; see also Townsend Hoopes, *The Limits of Intervention* (New York: David McKay, 1969), p. 202.

Chapter 6

92-93   This discussion of the early history of Soviet oil draws heavily on O'Connor, *World Crisis in Oil*, pp. 29-36, 45, 61, 79-90, 387.

93-94   This discussion of various Soviet energy sources draws on Robert W. Campbell, *The Economics of Soviet Oil and Gas* (Baltimore: Johns Hopkins University Press, 1968), pp. 1-15; data on current energy shares in Soviet Union are from the *New York Times*, April 24, 1974, p. 19 (firewood and all other noncommercial sources assumed by me unchanged from Campbell estimate for 1965).

94   This is the expert conclusion of Dimitri B. Shimkin, "Resource Development and Utilization in the Soviet Economy," in *Natural Resources and International Development*, Marion Clawson, ed. (Baltimore: Johns Hopkins University Press, 1964), p. 204. Dr. Shimkin is Professor of Anthropology and Geography at the University of Illinois.

95   U.S. and Soviet energy data for 1955 are from ibid., pp. 222-23.

95   On the implications of the Soviet emphasis on automobile and consumer durables, see Leo Huberman and Paul M. Sweezy, "Lessons of the Soviet Experience," *Monthly Review*, November 1967, pp. 14-18.

96   *Petroleum Press Service*, March 1973, p. 87.

96-99   For a fuller discussion of the two phases of Soviet foreign
             oil policy, see Tanzer, *The Political Economy of Inter-
             national Oil*, pp. 78-89.

97   National Petroleum Council, *Impact of Oil Exports from the
             Soviet Bloc*, vol. 1 (Washington, D.C.: National Petroleum
             Council, 1962), p. 41.

98   The Soviet view on the importance of the oil industry to
             the West is quoted from an "article in an authoritative
             Soviet publication," reported in ibid., p. 37.

99   For an analysis of the role of oil in relation to U.S.-Cuban-
             Soviet relations, see Tanzer, *The Political Economy of Oil*,
             pp. 327-40.

100   For the early history of China and oil, see O'Connor, *World
             Crisis in Oil*, pp. 37, 60.

100-102   This discussion of China and oil draws heavily on Tanzer,
             The Political Economy of Oil, pp. 277-87.

101   On China's alleged absence of oil, see the *New York Times*,
             January 5, 1974, p. 35.

102   *World Petroleum*, March 1961, p. 70.

102   Chinese oil data for 1961-1970 are United Nations esti-
             mates, in United Nations, *World Energy Supplies, 1961-
             1970*.

Chapter 7

108   Data on current energy shares in the underdeveloped
             countries are from ibid.

108-109   For a discussion of the efforts of the international oil
             companies, their home governments, and international
             organizations to block public oil exploration in the under-
             developed countries, see Tanzer, *The Political Economy of
             Oil*, pp. 90-106, 117-135.

109-110   Frankel, *Mattei*, p. 28; report of the United Nations
             Secretary-General on "Petroleum Needs of the Developing
             Countries," delivered to 2d sess., Committee on Natural
             Resources, January 3, 1972; U.N. Document E/C.7/20/
             Add. 2, point 7.

112   Data on the government-private share of Third World
             refinery capacity are estimated by the author for the end of

1972; "intermediate" refineries in small countries which import crude oil and process it primarily for export are excluded from the calculations.

112-13   Chase Manhattan Bank, *Capital Investments of the World Petroleum Industry: 1972* (New York: Chase Manhattan Bank, December 1973).

114-15   *New York Times*, April 28, 1966.

115-16   *Wall Street Journal*, April 16, 1974, p. 19.

Chapter 8

118   The Big Seven data for 1960-1969 are from First National City Bank, *Energy Memo*, October 1971, p. 4; OPEC data are from OPEC, *Annual Statistical Bulletin*, 1970.

118-19   A clear statement of Adelman's position at the end of 1969 is presented in " 'World Oil' and the Theory of Industrial Organization" (prepared at the end of 1969), reprinted in *Jackson Committee Hearings*, "Oil and Gas Import Issues: Part 3," Serial No. 93-3(92-38), January 10, 11, and 22, 1973 (Washington, D.C.: U.S. Government Printing Office, 1973), pp. 994-1011.

120-22   Many details of the Libyan oil events in the early 1970s are presented in *Church Committee Hearings, Part 5* (January 31, February 1, and February 6, 1974); on Exxon's turndown of Occidental, in particular, see pp. 79, 108-9.

121   Rifai, *The Pricing of Crude Oil*, pp. 308-9.

122-23   For this interpretation of the key role of the U.S. market and its oil import quota law, see ibid., pp. 310-23.

123   On the events surrounding the anti-trust restrictions at Teheran, see the testimony of James E. Akins of the Office of Fuels and Energy in the Department of State in *Church Committee Hearings, Part 5*, pp. 1-28.

123   The assistant Attorney General is quoted in the *New York Times*, June 6, 1974, p. 63.

124   Report on National District Attorney's Association, *New York Times*, June 4, 1974, p. 53.

124   On Irwin's role, see his testimony in *Church Committee Hearings, Part 5*, pp. 145-73.

124   Jack Anderson, *Washington Exposé* (Washington, D.C.: Public Affairs Press, 1967), p. 202; for specific examples, see pp. 202-5.

124-25   On Libyan oil nationalizations and the State Department, see *Wall Street Journal*, February 15, 1974, p. 1, and *Church Committee Hearings, Part 5*, pp. 29-58.

125   For oil cost and price estimates, see: for 1969, Adelman, *World Petroleum Market*, p. 209; for mid-1973, *The Economist* (London), July 7, 1973, Survey, p. 7.

128   The forecast of 1974 oil company pre-tax cash flow was made by the National City Bank of Minneapolis, January 8, 1974.

129   On the World Bank's estimate, see *The Petroleum Economist*, May 1974, p. 165; for an analysis of the bank's own negative role vis-à-vis the underdeveloped countries in the oil area, see Tanzer, *The Political Economy of Oil*, pp. 90-95.

129   On estimates of India's oil imports, see *Forbes*, June 1, 1974, p. 20.

133   On bilateral deals, see *Business Week*, March 2, 1974.

135   *Economist* (London), July 7, 1973, Survey, pp. 16, 19.

136-7   For an analysis of the fundamental factors leading toward a world depression, see Tanzer, *The Sick Society*, pp. 165-209.

137   Peter L. Bernstein, *New York Times*, June 9, 1974, section 3, p. 1.

Chapter 9

146   On the Chilean copper industry, see Frank, *Capitalism and Underdevelopment*, p. 61.

146   For a good discussion of the effects of "building on the best" on the underdeveloped countries, see John W. Gurley, "Maoist Economic Development: The New Man in the New China," *The Review of Radical Political Economics*, vol. 2, no. 4 (Fall 1970), pp. 27-28.

147   For a discussion of how research into coal liquefaction was blocked by the oil companies, see *Jackson Committee Hearings*, "Market Performance and Competition in the Petroleum Industry: Part 2," Serial No. 93-24 (92-59), November 28 and 29, 1973, pp. 755-60.

147   On the use of cattle manure, see the *Wall Street Journal*, March 5, 1974, p. 31.

148   On the Shah's investment in New York real estate, see
        Business Week, June 22, 1974, p. 52; through the Pahlavi
        Foundation, the Shah is known as the "first capitalist of
        the country" (ibid.).

151   Adam Smith, *The Wealth of Nations* (New York: Modern
        Library, 1937), p. 14.

Index

Abu Dhabi, 70
Aden (Yemen), 73-75
AGIP (Italian oil company), 84
Aiken (U.S. Senator), 33
Aldrich, Winthrop W., 40
Algeria, 110, 112; El Paso Natural Gas deal with, 35-36; French oil and, 80
Anderson, Robert B., 40
Anglo-Iranian Oil Company, *see* British Petroleum
Anglo-Persian Oil Company, *see* British Petroleum
Angola, 47
Arab Socialist League (Egypt), 75
Arabian Oil company, 88
Arabian American Oil Company (Aramco): founding, 26-27; U.S. foreign policy and, 40, 45
Atlantic Richfield Company, 25

Ba'ath Party (Syria), 75
Bahrein Island: U.S. oil companies on, 44-45
Bandung Conference (1955), 98
Batista, Fulgencio, 148
Bayonne, New Jersey (Exxon refinery), 100
Big Seven (Seven Sisters), *see* International oil companies
Bolivia, U.S. oil interests in, 47-48

Breeder reactors, 20
British Petroleum (Anglo-Iranian Oil Company and Anglo-Persian Oil Company), 17, 49, 51, 82; British government and, 41-44; international growth of, 26, 27; Kuwait oil and, 46; Turkish Petroleum Company and, 79
Buckley, John, 83
Burmah Oil Company, 43

Capitalism: "energy crisis" and, 151-52; irrational energy policy of, 143-52
Caltex (Standard Oil of California and Texaco), 101; in Saudi Arabia, 45
Cattle manure, as fertilizer, 147
Cefis, Eugenio, 86
CENTO alliance, 65
Central Intelligence Agency, 60; Iranian coup and (1953), 50-51
CFP, *see* Compagnie Française des Pétroles
Chase Manhattan Bank, 112
Chile, copper industry in, 146
China: British imperialism and, 145; coal reserves in, 21; oil development in, 113-15; oil industry in, 99-103